THE
UNIVERSE
EXPLAINED
WITH A
COOKIE

TO MY DAD, WHO SHOWED ME HOW WE ARE CONNECTED TO EVERYTHING.
TO MY MOM, WHO SHOWED ME HOW EVERYTHING IS CONNECTED TO US. —G.E.

FOR LIZZIE AND JIM, MY FAVORITE COOKIE EATERS.
NEVER STOP ASKING QUESTIONS. —M.K.

JOYFUL BOOKS FOR CURIOUS MINDS

An imprint of Macmillan Children's Publishing Group, LLC
Odd Dot® is a registered trademark of Macmillan Publishing Group, LLC
120 Broadway, New York, NY 10271 • OddDot.com • mackids.com

Molasses Disaster site photo by Globe Newspaper Co., courtesy of Boston Public Library.

EDITOR Julia Sooy
DESIGNER Tim Hall
PRODUCTION EDITOR Kristen Stedman
PRODUCTION MANAGER Jocelyn O'Dowd

Library of Congress Control Number 2023029173
ISBN 978-1-250-83039-5

Our books may be purchased in bulk for promotional, educational,
or business use. Please contact your local bookseller or the Macmillan
Corporate and Premium Sales Department at (800) 221-7945 ext. 5442
or by email at MacmillanSpecialMarkets@macmillan.com.

First edition, 2024
Printed in China by 1010 Printing International Limited, Kwun Tong, Hong Kong

1 3 5 7 9 10 8 6 4 2

THE
UNIVERSE
EXPLAINED
WITH A
COOKIE

WHAT BAKING COOKIES CAN TEACH US ABOUT QUANTUM MECHANICS, COSMOLOGY, EVOLUTION, CHAOS, COMPLEXITY, AND MORE

GEOFF ENGELSTEIN

ILLUSTRATIONS BY MICHAEL KORFHAGE

NEW YORK

CONTENTS

INTRODUCTION

This is a chocolate chip cookie:

It's about two inches across.

This is our galaxy,
the Milky Way:

It's about 40,000,000,000,000,000,000,000 inches across.

Quite a difference, right?

But by exploring what goes into the cookie—the ingredients and the steps—we can learn an awful lot about how everything works, from the tiny world of subatomic particles to galactic clusters.

We think there are about 10^{12} galaxies in the universe. And most galaxies have about 10^{12} stars. So there are around 10^{24} stars in the universe.

As a shorthand, we will write really large and small numbers using exponential notation—the first digit, and the number of zeros. So the size of the Milky Way in inches would be written as 4×10^{22}. That's four followed by twenty-two zeros.

For small numbers, the exponent is negative. The diameter of a proton is 0.00000000000003 inches. That's thirteen zeros and a three. That's written as 3×10^{-14} since you have to move the decimal fourteen places to the right until it's just after the three.

One gram of the stuff that makes up most plants and animals has about 10^{23} atoms. A cookie is very similar in ingredients, and weighs about 16 grams. So a cookie has…about 10^{24} atoms.

The number of atoms in a cookie is about the same as the number of stars in the universe.

Science is about making observations, and connecting them to other observations and ideas, to try to understand how things work, how we can do things better, and make predictions about what might happen in the future. And all those observations started with something every day that we encounter as humans—waterfalls, plants and animals growing, wind and weather. While science often seems like an abstract and obtuse subject, it all starts with the commonplace.

And a cookie is a great place to start. The ingredients connect to a wide variety of science. And at the end you get a tasty treat.

Each chapter in this book starts with an ingredient or step used to bake a cookie, or a topic that is cookie-adjacent. We then build out from there to explore a variety of ideas in science. This book is about drawing connections between things you might not have thought of as being connected.

Here is my mom's recipe for chocolate chip cookies:

CHOCOLATE CHIP COOKIES

INGREDIENTS

2 ¼ cups all-purpose flour

1 teaspoon baking soda

1 ¼ teaspoon salt

1 cup (2 sticks) butter, softened

¾ cup granulated sugar

¾ cup packed brown sugar

1 teaspoon vanilla extract

2 large eggs

12 ounces chocolate chips

Preheat oven to 375°F.

Use a fork to stir flour, baking soda, and salt together in a bowl. Use a hand mixer or stand mixer to beat butter, granulated sugar, brown sugar, and vanilla extract until creamy. Add one egg at a time, beating well. Pour in flour mixture, continuously mixing. Fold in chocolate chips. Scoop cookie dough onto baking sheets lined with parchment paper.

Bake for about 10 minutes or until golden brown.

LET'S GET BAKING!

CHAPTER ONE

DARK MATTER EXPLAINED WITH FLOUR

About 85 percent of the matter in the universe is missing. We've given it a name—*dark matter*—but haven't captured it, produced it in a lab, or directly sensed it in any way. So how do we know it's there? To answer that question, let's turn our attention to flour and dough.

One of the fun parts about making dough is the typical springy, spongy texture. Kneading it, bouncing it, stretching it, and shaping it—all give a visceral pleasure.

Flour gives that stretchy feel to the dough, thanks to proteins it contains called *glutens* (yes, these are the same glutens that cause digestive and other issues with some people). Glutens act like tiny springs that hold together the starch and other particles in the dough. As the dough is pulled, the gluten molecules stretch and pull it back together.

They apply a *force* that holds the dough together.

We know of only four fundamental forces in the universe: gravity, electromagnetism, the strong force, and the weak force. Basically:

* **Gravity** is the force of attraction between all objects in the universe. It makes planets orbit the sun and cookies fall when you drop them.

* **Electromagnetism** is the force between positive and negative charges. It is the force behind electricity and light, and it binds atoms and molecules.

* The **strong force** holds protons and neutrons together in atomic nuclei.

* The **weak force** is involved in radiation and other subatomic processes.

Can you guess which force is the strongest? Yes, it is the strong force. Although the weak force is not the weakest—that dubious honor goes to gravity. Physicists are not great at naming things—look forward to more bad names in future chapters.

But the strong force is the most powerful of the forces—over a hundred times stronger than electromagnetism, a million times stronger than the weak force, and 10^{38} times stronger than gravity. But both the strong and weak forces only work over very short distances—around the size of the nucleus of an atom. So we do not notice them in our normal lives.

The forces we deal with every day are electromagnetism and gravity. The force that the gluten molecules create to hold the dough together is an electromagnetic force. It is caused by the negatively charged electrons and positively charged protons in the gluten protein attracting and repelling one another in different ways. The shape of a molecule is determined by different parts of the protein pulling or pushing one another. When it gets twisted or stretched, it wants to return to the original shape, much as a stretched spring wants to pull back together. We'll talk about the shapes of molecules a lot more when we get to baking soda in Chapter Three, but the key point is that these shapes are caused by the electromagnetic force.

The electromagnetic force is much stronger than gravity. To demonstrate that, go get a cookie and put it on a table. The entire Earth—all 7,000,000,000,000,000,000,000 tons of it—is pulling on the cookie through the force of gravity, holding it to the table and preventing it from floating away.

Now grab it with your hand, and lift it off the table. The electromagnetic force generated by the muscles in your arm is easily stronger than all the gravity generated by the entire planet beneath you. Take that, Earth!

But when astronomers look at the forces that affect the solar system and galaxies, they almost never have to consider the effects of electromagnetism—just gravity. Why is that?

The answer lies in the different way the forces operate. Electromagnetic force is created when you separate a positive and negative charge. The force wants to pull them back together. If you hold a negatively charged electron and positively charged proton near each other, they will want to snap together, similar to the north and south poles of a magnet.

But once the electron and proton are together, there is no more electromagnetic force beyond them. They become neutral, bound up in a little package—a hydrogen atom, in this case. Over time, positive and negative charges attract one another and cancel out the electromagnetic force.

Gravity, on the other hand, doesn't get canceled out. Every particle creates gravity, and it all adds together. In essence, whereas there are two "flavors" of electromagnetism—positive and negative—there is only one flavor of gravity. Despite what you see in science fiction movies and books, we have not figured out a way to make antigravity. We don't even have a theoretical way to do it. Gravity is gravity.

So, unlike electromagnetism, all the gravity in the universe just keeps adding up. There's nothing to cancel it out. When you get to huge masses, such as stars and black holes and galaxies, and huge distances, gravity is the only game in town.

Over short distances in space, you can certainly get electromagnetic disturbances. For example, the sun emits a stream of electrons and protons called the solar wind, which has a lot of electromagnetic energy. It causes the aurora—the northern and southern lights—as those charge particles hit the earth's magnetic field.

But when looking at stars in a galaxy, gravity is orchestrating their motion.

Galaxies rotate. Like anything that rotates, a force must be pulling the stars toward the center to keep them rotating, rather than flying off into space.

Let's put our cookie dough aside for a second and pick up another type of dough—pizza dough. Like cookie dough, pizza dough is mainly held together by glutens.

If you've seen pizza dough thrown into the air and spun around, it spreads out because the other parts of the dough want to move in a straight line. The glutens have to stretch out to exert force to keep the dough moving in a circle. If I spin it too quickly, the glutens won't exert enough force and the dough will tear and fly apart. And most likely make a mess.

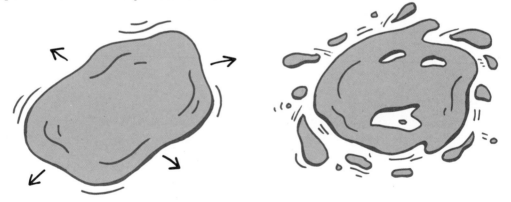

Gravity is the gluten of the galaxy. It holds the whole thing together. So the force that gravity can exert controls how fast the stars can rotate.

In the late 1970s, Dr. Vera Rubin did a detailed observation of how quickly the stars in galaxies were rotating around the center. What she and everyone else expected was that the farther a star was from the center of the galaxy, the slower it would rotate. That's true for our planets around the sun—the Earth takes one year for the trip, while Saturn takes twenty-nine years. The winds in a hurricane are slow around the edges and get faster as you move closer to the eye.

But her measurements showed that the stars on the edges of galaxies were moving way faster than they should. Galaxies should be flying apart. What mysterious force was holding them together?

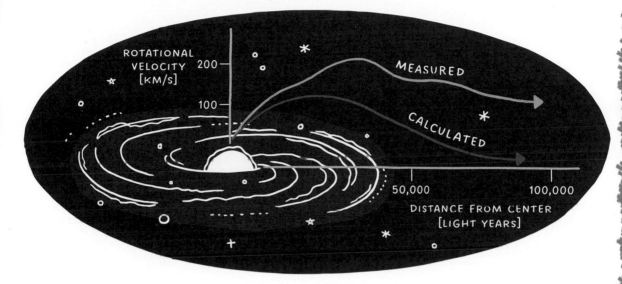

ACTUAL STAR ROTATION SPEEDS VS WHAT IS EXPECTED.
THE OUTER STARS SHOULD FLY RIGHT OUT OF THE GALAXY.

When dealing with an observation that doesn't fit what the rules of science—the theory and equations—predict, there are two options. The first is to assume that the rules are correct and that there is some problem with the observation. Maybe there's an error in the way the measurement was done. Or perhaps there's extra matter we didn't know about that was keeping those stars in place with extra gravity.

The other possibility is that the theory and the equations are not correct. Perhaps they work under some situations, but not the conditions we are looking at. The laws of gravity that Einstein developed were based on observations of stars and planets. Maybe the rules for gravity change slightly over very large distances. Perhaps it doesn't drop off quite as quickly as we think, so gravity is stronger at the edge of the galaxy than we expect.

When presented with a dilemma like this, it is usually simplest to try to keep the rules the same and figure out what could be happening under those rules. After Rubin and her team published their measurements, a series of other experiments verified those results and found that other galaxies were doing the same thing.

The simplest explanation is that space contains a type of matter that we don't know about and is very difficult to detect, since we haven't seen it before. It must be invisible and not interact with light at all, since light just passes right through it. So the mysterious substance was named *dark matter*. Dark matter was providing extra gravitational gluten to hold the galaxy together, allowing the outer-rim stars to zip along.

In the fifty years since we have been trying to figure out what dark matter consists of, we still have no idea. We've gradually been ruling out possibilities and have a few clues to go on:

* It doesn't interact with light at any frequency we've looked at, from radio waves to x-rays.
* It doesn't clump in the center of galaxies as much as regular matter. It likes to spread out.
* It makes up 85 percent of the mass in the universe.

If the dark matter theory is correct, we only know what 15 percent of the universe is made of! 85 percent of the mass is a complete mystery. There have been many proposals to tweak general relativity, our best explanation of gravity, so that dark matter is not needed—so that the gravity is a bit stronger in the right ways to explain the motion of galaxies. Unfortunately, recent scientific observations have undercut many of these ideas.

So, for now, we think dark matter is the gluten of galaxies, binding stars together.

CHAPTER TWO

FUSION EXPLAINED WITH SUGAR

To create the forces we discussed in the last chapter, we need *energy*. Energy comes in many forms—movement, heat, electricity, chemical bonds, and more. The advance of civilization has gone hand in hand with our ability to generate energy more efficiently. Wood, coal, oil, and nuclear power each get more energy from the same amount of stuff, but all also come with a cost for our planet. Fusion power could solve all these problems.

All types of energy are related and can be converted into one another. So let's start by looking at the key source of energy for humans and animals, and one of our cookie ingredients—sugar. There are actually many types of sugar. The one we cook with is called *sucrose* and is made up of two smaller sugar molecules joined together—*fructose* and *glucose*.

When we eat a cookie, our body breaks down the sucrose and other carbo-hydrates, converting them to a form the body can use. They get stored in *glycogen* in muscles and the liver for short-term use, and in *triglycerides* in fatty tissues for long-term storage.

Blood carries glucose to the cells in your body. On average, about 4 grams of sugar are in your bloodstream—basically a teaspoon. A teaspoon of sugar helps the energy go round.

The sugar is absorbed by cells and transported to the *mitochondria* inside them that produce energy. The glucose goes through a series of chemical reac-tions and combines with oxygen, which releases the energy that keeps living things alive.

How much energy do we get from a cookie? Fortunately, there's a handy guide right on the side of the box—the calories. Calories measure how much energy a food can give to your body. There are other measures of energy besides calories—horsepower, BTUs, kilowatt-hours, joules, and many more. But they all measure the same thing, just as you can measure distance with inches, feet, centimeters, miles, furlongs, or light-years. You can use any unit you want, but measuring the distance from New York to Boston in inches or light-years isn't exactly practical.

My in-depth research for this book involved reading nutrition labels for a lot of different types of cookies. And eating them, of course. For science. So I can tell you that the average cookie is about 150 calories. When that energy is released, it is enough to power a standard household 10-watt LED bulb for about seventeen hours.

Maybe the witch in *Hansel and Gretel* keeps the lights on by breaking off pieces of gingerbread.

Our body takes the energy out of the cookie and uses it to power our cells. But how does the energy get into the cookie in the first place? Well, the sucrose we put into the cookie dough is extracted from plants like sugarcane and sugar beets, so the energy came from those plants. But how did those plants get the energy to put into the sugar? They use a process you've probably heard of: *photosynthesis.* Photosynthesis is a series of chemical reactions that combine sunlight, carbon dioxide, and water, turning them into sugar and oxygen.

Ultimately, the energy we get from eating a cookie comes from the sun. In fact, just about all energy on earth originates with the sun. The main exceptions are geothermal energy, which comes from heating within the earth itself, and tidal energy, which comes from the interaction between the earth and the moon. But basically, everything on our planet is powered by the sun, even some things you might not realize.

For example, hydroelectric power, where water builds up behind a dam and spins turbines as it flows, would seem to get energy from gravity. It is gravity that pulls the water from the mountains to the oceans, converting potential energy into kinetic energy.

But once the water reaches the ocean, it needs to get back to the mountaintop, or else our dams will quickly run dry. That happens by water evaporating from the ocean, forming clouds, and falling back to higher ground in the form of rain or snow. It is energy from the sun that powers that evaporation.

Sunlight provides the vast majority of the energy that drives our planet, yet we only receive a small piece of the sun's power—the slice that hits our planet is less than a *billionth* of the energy it puts out.

How does the sun create so much energy that only a billionth of it is more than enough to provide energy to every living thing on earth? It uses the strongest of the forces we discussed last chapter—the aptly named strong force, which holds together the protons and neutrons that lie at the center of the atom.

When you bring two nuclei together, at first, they want to push each other away. The nuclei are positively charged, and when they get very close together, the charges repel each other, the same way you feel a force when you try to push together two north ends of magnets. Both the magnets and nuclei pushing apart are from the electromagnetic force.

But if you push them close enough together, the nuclei will want to bind together, thanks to the strong force. The strong force is what holds protons and neutrons together in the nucleus. And while it is stronger than the electromagnetic force, it only works at very short distances. So it doesn't help pull the nuclei together until they are very, very close. You need to supply a lot of energy to overcome the electromagnetic force and push the nuclei close enough together for the strong force to kick in. But once you do, they release energy—a lot of energy. This is called *nuclear fusion*. And the energy released is way more than what you need to use to move the nuclei close enough for them to fuse. So you put a lot of energy in, but you get even more out.

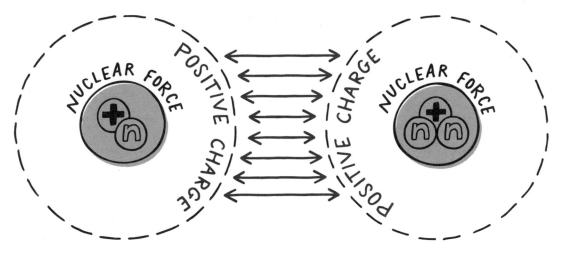

THE ELECTROMAGNETIC FORCE PUSHES THE NUCLEI (BLUE PARTICLES) APART UNTIL THEY GET VERY, VERY CLOSE. THEN THE STRONG FORCE TAKES OVER AND PULLS THEM TOGETHER.

Actually, it is lighter nuclei that release energy when they are fused together. Hydrogen (one proton), deuterium (one proton and one neutron), and tritium (one proton and two neutrons) are among the nuclei that release the most energy. Once you get to iron or heavier atoms, you don't get any energy from fusion anymore. And if you go to very heavy atoms, you get energy by splitting up the nuclei, instead of joining them together. This is called *nuclear fission*.

The heavier the atoms are, the more fission energy they produce—and uranium and plutonium are the heaviest ones around. The energy from fission is also from the strong force, and it is what powers nuclear reactors and nuclear weapons.

Our sun, like all stars, is made up mostly of these lighter nuclei, giving them a lot of fuel to power nuclear fusion. But how does the sun overcome the energy barrier to force nuclei close enough together for fusion to take place? Gravity.

The sun has about 330,000 times as much mass as the earth. ("Mass" is a measure of how much stuff something is made of.) Gravity pulls all that together, squeezing it at the core and raising the temperature. The high temperature and pressure make it possible for energetic nuclei zipping around to smash into one another fast enough to overcome the electromagnetic repulsion, fuse together, and release energy. This energy helps increase the temperature inside the star even more, leading to a chain fusion reaction.

Jupiter, the largest planet in our solar system, is made up of similar materials as the sun. But it doesn't have the mass required for gravity to squeeze it together enough to start a fusion reaction. It would need to be about eighty times as large. But if it was, we would live in a binary star system.

Fusion creates a lot of energy. But how much, compared with other sources? Like, say, the sugar in a cookie?

Our chocolate chip cookie has about 150 calories. The atoms in it are not great for fusion, but let's say we made a cookie completely of particles good for fusion, like hydrogen and deuterium. If we fuse all the stuff in that not-great-tasting cookie, it would release about *8.2 billion* calories. Not so good if you're on a diet, but really good if you're looking to, say, power a country.

Can we create fusion reactions on earth?

We have actually done it. Hydrogen bombs are powered by fusion. But to make those work, we need a fission explosion with heavy nuclei like uranium and plutonium, and we need to force the explosion inward to compress the deuterium and tritium core, which kick-starts the fusion reaction. That we need to explode an atom bomb *inward* to start a fusion reaction should give you some idea of the incredible temperatures and pressures it needs to get kick-started.

Exploding atomic bombs isn't practical for power generation for obvious reasons. But it's not easy to create the temperatures and pressures at the center of a star here on earth in a controlled way. They are so intense that no physical container could hold them. We've been trying for over fifty years to figure out how to do it, and in 2022 we finally did an experiment that reached "break even"—where the power we get out is at least equal to the power we put in. But we still have a lot of work to do to make this a practical process for providing power.

There are two main methods researchers are trying to create fusion power reactors. The first is to suspend the fuel in intense magnetic fields and heat it. The electromagnetic force keeps the plasma fuel ("plasma" is a state where the temperatures are so high that electrons are no longer bound to the atomic nucleus) contained and compressed as it heats up to immense temperatures. As you can imagine, this is not a simple task.

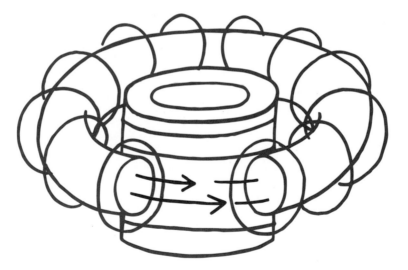

THIS FIRST METHOD IS ACCOMPLISHED
IN A MACHINE CALLED A TOKAMAK

The other method is to fire high-power lasers from multiple directions at a single fuel pellet. The lasers heat up and compress the pellet enough to initiate fusion and release more energy than required to power the lasers. The idea for a reactor is that pellets would be dropped into the reactor one after the other, with the lasers turning on and blasting each one after it drops into place.

Fusion power has always been one of those technologies that seem like it is tantalizingly close but just out of reach. But if we can build reliable fusion reactors, it will transform the planet. Fusion runs on plentiful fuel that is found in seawater, and it doesn't leave radioactive waste like nuclear reactors.

Since the strong force is the strongest of forces, is that the best we can do to generate energy? Is fusion the pinnacle?

There is one more potential step, thanks to Albert Einstein.

As a consequence of his theory of relativity, Einstein showed that energy and matter are interchangeable. Energy has mass, and mass can be converted into energy. They are linked by perhaps the most famous equation in all of science:

$$E = mc^2$$

The "E" stands for energy, "m" for mass, and "c" the speed of light.

The speed of light is very, very fast—186,000 miles per second—so when you square it (multiply it by itself), you have a really large number. This means that a small amount of mass is equivalent to a huge amount of energy.

If we could convert mass directly to energy, we would release even more than we get from a fusion reaction.

Back to our cookie. If we change all the mass in a typical cookie to energy, we get about 430 trillion calories—way more than the eight billion from our fusion reaction, and the paltry 150 calories from the chemical energy in the sugar and other ingredients.

430 trillion calories is about 500 gigawatt-hours of power—enough to provide 500 billion watts for an hour—or enough to provide the entire energy requirements of the United States for about an hour.

Pretty good for just one cookie.

Unfortunately, the only way we know of to completely liberate all the energy contained in matter is to collide it with *antimatter*. Every particle has an *antiparticle* partner. If the two meet, they annihilate into a burst of energy. Antimatter is produced in high-energy particle collisions, but it typically doesn't last long and is hard to store. You can't keep it in a container since it will annihilate when it hits the edge of the container. As with fusion plasma, we need to use magnetic fields in vacuums to contain antimatter and keep it from regular matter so we can study it.

So, regretfully for science fiction stories, unless we can find some hidden source of antimatter in the universe, antimatter cookies are not the future of power.

THE MYTHICAL COOKIE-POWERED ANTIMATTER STARSHIP DRIVE

CHAPTER THREE

ATOMIC STRUCTURE EXPLAINED WITH SALT AND BAKING SODA

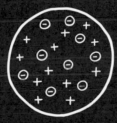

Everything around us—like this book—is made of molecules. Salt, for example, is made up of two atoms—a sodium (Na) and a chlorine (Cl). Here's a diagram showing the molecule:

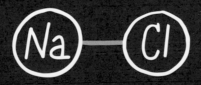

The line between the two atoms is called a *bond*. Bonds hold molecules together. It basically means the atoms share an electron. Both the atoms that make up molecules and the way they are arranged determine how hard the material is, the color, what it likes to combine with, and other properties. The way the atoms are attached to one another can be more important than most people realize. Diamonds and graphite (pencil lead) are both made of the same atoms—carbon. The difference lies in the way they are connected to one another.

Baking soda—also known as *sodium bicarbonate*—is a little more complex than salt. It has one sodium (Na), one hydrogen (H), one carbon (C), and three oxygens (O). Here's the diagram:

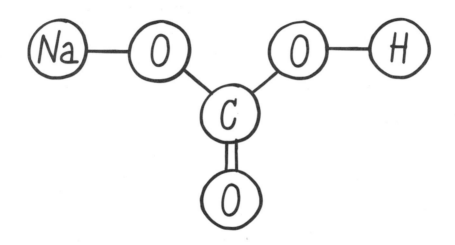

See that double line going from the carbon to the oxygen at the bottom? That's called a double bond—atoms sharing two electrons. To understand bonds, let's look at something you have probably seen before—the periodic table of elements.

In this table, atoms (which were called *elements* when the table was created—hence the name) are listed in order of the number of protons they have (a standard neutral atom always has the same number of electrons and protons).

Hydrogen is the simplest, with one electron and one proton. Carbon has six of each, and oxygen eight. Each box on the same row, from left to right, has one more proton and one more electron. When you reach the last element on the right, the next bigger atom will be the first box on the next row.

Here are the first five rows of the periodic table:

Each column has similar chemical properties. That's what helped Dmitri Mendeleev, the originator of the table, to assemble it in the first place. But the last column, from helium on top to xenon on the bottom, is perhaps the most interesting. These are called the noble gases, and they are basically inert. They do not react with other chemicals, which made them very difficult to detect.

In fact, none of the noble gases were discovered until decades after the periodic table. Argon was the first. It was discovered accidentally by experiments performed by Lord Rayleigh in the 1890s. He was examining air by removing all the component gases, one by one, mainly to get detailed measurements on nitrogen. However, he was always left with a small bubble of gas that he could not get to react with anything. William Ramsay suggested that it might be a new element, and together they did experiments proving that was the case. They named the new element argon, from the Greek word for "lazy"—which seems a little judgmental.

It turns out that 1 percent of the atmosphere is made of argon, but because it doesn't react with anything, we never knew about it. Shortly thereafter, neon and the rest of the noble gases were discovered, and the final column of the periodic table was completed.

This column is arguably the most important one in the entire table. These element numbers—2, 10, 18, 36, and 54—make for happy atoms. Those numbers are what all atoms want to be. That's why they don't react with other elements. They don't want to share electrons since they have the perfect amount.

Chlorine, for example, is 17. If it had just one more electron, it would have 18, one of our happy numbers. Sodium, on the other hand, has 11. If it had just one less electron, it would have 10, another happy number.

Sodium has one electron it wants to get rid of, and chlorine would really like one. How about if they get together and sodium lets chlorine borrow an electron? That's what a chemical bond is—atoms sharing electrons. And that's how our salt molecule, NaCl, gets made.

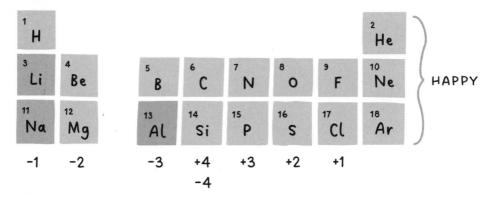

Here are the first three rows of the periodic table, with the columns labeled for how far away they are from being happy. The atoms in the first column all have one more electron than the last column. Beryllium and manganese have two more, and oxygen and sulfur have two fewer. Carbon and silicon are right in the middle—so they can be considered either +4 or −4.

Take another look at baking soda: ⟶

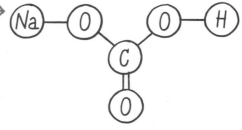

Notice that the number of lines coming out of each atom match how far the column is from the happy column. The sodium and hydrogen atoms each have one line, oxygen has two, and carbon has four.

Let's get back to cookies. Why is baking soda added to our dough? What is it doing?

Bonds between atoms are not all created equal. Some are stronger than others. If an atom is given a better offer—exposed to atoms or molecules that it would rather bond to, or where the bonds would be stronger—it will jump ship and make a new molecule.

The bonds in baking soda are not all that strong. In particular, the bond between the sodium and oxygen atoms is particularly weak.

The atoms in baking soda would much rather bond with a hydrogen atom than the sodium. If it is placed in an area with extra hydrogen, it will break down and recombine into three pieces: a free sodium atom, a water molecule, and a carbon dioxide molecule.

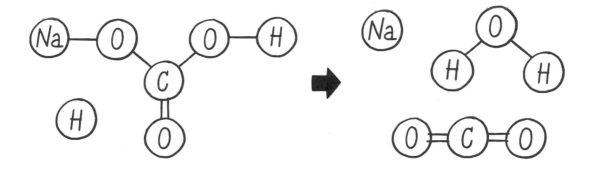

As you can see, the atoms on the left match the atoms on the right. They just rearranged their connections.

Carbon dioxide is a gas, which fluffs up the cookie, making it light and airy.

Where does the extra hydrogen atom come from to make this reaction? Some substances tend to release extra hydrogen—you know them as *acids*.

Many molecular bonds break down in the presence of hydrogen, and they re-form into simple forms like the water and carbon dioxide that come out of baking soda. This is why strong acids like sulfuric acid and hydrochloric acid can be so dangerous, particularly to the types of molecules that form our bodies. The extra protons released by acids will break down organic molecules.

We don't cook with serious acids, of course. However, many mild acids we use can trigger reactions like this, like vinegar, lemon juice, or yogurt.

You may have seen the classic science fair "baking soda volcano." Combining baking soda and vinegar rapidly creates carbon dioxide and water, leading to a foamy, bubbling, and very dramatic "explosion." A little red food coloring mixed in, and you've got a nice lava flow!

The periodic table does a good job of explaining how atoms attach to one another. But what's the recipe for the atom itself? In the 1890s, British physicist J. J. Thomson showed that atoms included a particle that carried a negative electric charge that was much, much lighter than the full atom. Also, he found that the particles were the same regardless of which element they came from. He called them *corpuscles*, which, honestly, is not the best name. Fortunately, other scientists adopted the term *electrons*.

With the discovery of the electron, Thomson proposed a model of the atom called the *plum pudding* model. Plum pudding is a traditional British dessert, made with plums, raisins, and other fruit—a relative of the American fruitcake.

* * *

He suggested that atoms were composed of a ball of positive stuff (the cake) with electrons floating around inside (the raisins).

I prefer to think of it as an M&M cookie, with the electrons represented as the candy bits.

The number of M&Ms in our neutral cookie determines which element it is. Hydrogen has one, carbon has six, and oxygen has eight.

This model made sense, and scientists did experiments to understand more about the positive dough part of the metaphorical cookie. What was it exactly?

In 1909, Ernest Marsden and Hans Geiger (who would later go on to invent the Geiger counter to measure radiation) did an experiment under the direction of Ernest Rutherford. They shot helium nuclei (then called *alpha particles*) at a very thin sheet of gold foil. They expected that they would have their paths slightly bent as they moved through the foil, but basically pass through. They should easily punch through the "cookie" part of the M&M cookie, so, at first, they just set up detectors on the far side of the gold foil. While most of the particles passed through as they expected, they weren't detecting every particle they shot. Some were disappearing. Maybe they were getting absorbed? Or perhaps ending up somewhere else?

Geiger and Marsden kept moving their detectors farther and farther away from the "straight through" path and found that while most of the alpha particles were passing through the foil, some were bouncing off at crazy angles.

Rutherford was astonished. He later wrote:

> It was quite the most incredible event that has ever happened to me in my life. It was almost as incredible as if you fired a 15-inch shell at a piece of tissue paper and it came back and hit you.

The theoretical "cookie dough" part of our M&M cookie atoms should not have been dense enough for the alpha particle bullets to bounce off. And they couldn't be bouncing off the candies—the electrons. The alpha particles were ten thousand times as massive as the electrons. They'd just blow right past them.

He did some math and realized that the only model that matched the results was if atoms were mostly empty space, with a very, very concentrated, positively charged nucleus and electrons hovering around it. A small fraction of the alpha particles was hitting this nucleus and bouncing off it in all directions, like a Ping-Pong ball hitting a billiard ball.

The atom couldn't be a low-density cake taking up the full volume of the atom. All the cake material was squished into a very, very tiny area in the middle of the atom. The plum pudding model was dead.

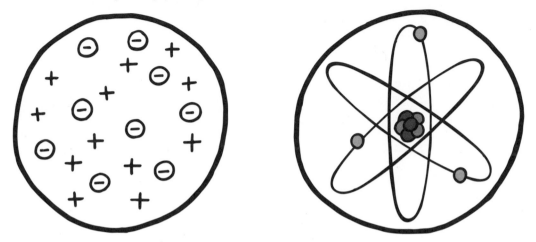

The classic picture on the right side of the electron "orbiting" the atom's nucleus like planets orbit the sun is actually very misleading. Electrons don't really move in circles. They flit from place to place, creating an electron cloud around the atom. This is due to quantum mechanics—the physics of very tiny things—which we will dive into in a later chapter. So while this orbiting picture is a better model than the plum pudding model, it's still far from the perfect picture. Here's a slightly better one:

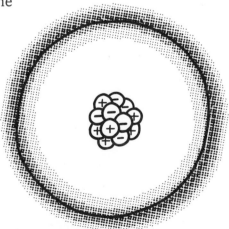

The scale of atoms—the size of the nucleus versus the size of the atom overall when you include the electron cloud—is not really comprehended by most people. The standard picture of an atom does not give you a clue as to how small the nucleus is.

As a way to picture this, let's blow up an atom until the nucleus is the size of a chocolate chip. How far away will the electron cloud be? Any guesses?

Let's put our chocolate chip nucleus on the fifty-yard line of a football field—right smack in the middle. The electron cloud, which defines the size of the overall atom, will be *at the end zones.*

Sometimes you'll hear people say that atoms are mostly empty space—this is why. Everything between the chocolate chip at the fifty-yard line and the electron cloud that starts around the goal lines is empty. That's *a lot* of empty space.

CHAPTER FOUR

QUARKS EXPLAINED WITH A COOKIE SWAP

Exchanging cookies for holidays and other celebrations is a fun tradition and a great way to discover new cookies and great recipes. Let's say you have a friend who throws a "blind cookie exchange." Everyone brings their cookies wrapped up nicely in little bags. Unfortunately, we can't see inside them, and the etiquette at this party doesn't allow us to peek. How can we find out what cookies we've gotten?

Well, if you're an experimental physicist, your first instinct is to fire two of these cookie bags at each other really, really fast and then sift through the debris to see if you can figure out what the cookies were.

After coming up with the nice, neat little model of electrons, protons, and neutrons, physicists wondered what would happen if protons, well, smashed into each other very, very hard. Were they composed of even smaller parts?

When this experiment was done, none of the protons broke down into smaller pieces that could be detected. However, a whole bunch of new types of particles were created, all roughly the same mass or larger. It was as if we smashed together bags of cookies, but all we got back were bags of different cookies—but still bags of cookies.

Eventually, by examining all these new members of the "subatomic zoo," as it was called, physicists came up with a model for what was going on inside the proton and neutron—called the quark model.

Let's go back to the bags of cookies from our cookie swap. Each bag represents a type of particle like a proton or neutron (let's leave electrons to the side for now), which are generally called *baryons*. All baryons have three cookies in their bag. They come in different flavors—chocolate chip, oatmeal, sugar, macadamia nut, and others. And each flavor comes in three colors—red, blue, or green. So there are red oatmeal, blue oatmeal, and green oatmeal, red sugar, blue sugar, and green sugar, and so on.

FOR OUR PURPOSES, WE'LL USE RED , BLUE , AND GREEN

RED
CHOCOLATE CHIP

RED
OATMEAL

RED
SUGAR

RED
MACADAMIA NUT

BLUE
CHOCOLATE CHIP

BLUE
OATMEAL

BLUE
SUGAR

BLUE
MACADAMIA NUT

GREEN
CHOCOLATE CHIP

GREEN
OATMEAL

GREEN
SUGAR

GREEN
MACADAMIA NUT

Quark theory says that:

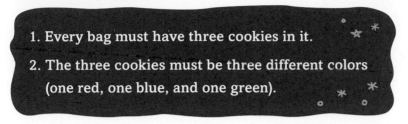

1. Every bag must have three cookies in it.

2. The three cookies must be three different colors (one red, one blue, and one green).

As long as you follow those rules, you can have any combination of cookies. So, for example, you could have:

* RED CHOCOLATE CHIP, BLUE CHOCOLATE CHIP, GREEN OATMEAL.

* RED OATMEAL, BLUE SUGAR, GREEN MACADAMIA NUT.

* RED MACADAMIA NUT, BLUE SUGAR, GREEN CHOCOLATE CHIP.

Any of those would make up a valid bag.

There's one other cookie-bag rule in quark theory:

3. Bags act differently based on the three flavors in the bag, not their colors.

For example, maybe oatmeal cookies are heavier than sugar cookies. So a bag with three sugar cookies wouldn't weigh as much as a bag with one sugar cookie and two oatmeal cookies. But these two bags...

* RED OATMEAL, BLUE SUGAR, GREEN SUGAR

* RED SUGAR, BLUE SUGAR, GREEN OATMEAL

...would behave exactly the same. They would have the same weight and any other properties. You couldn't tell them apart just by measuring the bag.

In quark theory, the "bag" represents the particle, like a proton or neutron. The cookies inside the bags are called *quarks*.

Like the cookies, quarks can come in multiple flavors. We currently know of only six, and they come in three pairs, with each successive pair being heavier. The two lightest are called Up and Down. The middle-weight quarks are called Strange and Charm, and finally the two heaviest quarks we are aware of are called Top and Bottom. Originally those were named Truth and Beauty, but physicists decided they had already gone a little too far with Strange and Charm and wanted to make their theory sound more serious again.

In our cookie-bag model, Up, Down, Strange, Charm, Top, and Bottom are the different flavors of the cookies. And each of these six flavors also comes in three colors—red, blue, and green.

A proton is made of two Up quarks and a Down quark (UUD). A neutron is one Up and two Downs (UDD). But those aren't the only combinations, of course. Each unique combination of quarks results in a particle with different characteristics, a recipe if you will. Here are a few of the combinations, along with the increasingly exotic names physicists have given to them:

SIGMA = UP UP STRANGE

LAMBDA = UP DOWN STRANGE

CHARM XI PRIME = UP STRANGE CHARM

DOUBLE CHARM BOTTOM OMEGA = CHARM CHARM BOTTOM

Other than protons (UUD) and neutrons (UDD), all the other combinations of quark flavors have an incredibly short lifetime—trillionths of a second or shorter. But we can see them in particle detectors and learn about their properties.

While thinking about these particles as bags of three cookies helps to visualize what's going on, reality is obviously much more complicated.

First, no bag holds the cookies together. They're not inside anything at all. Instead, special particles called *gluons* help glue them together. Perhaps the name is a bit too on the nose, but it does get the point across. Gluons transmit one of the four fundamental forces we talked about in Chapter One—the strong force. You can think of them as tiny springs that run between the quarks.

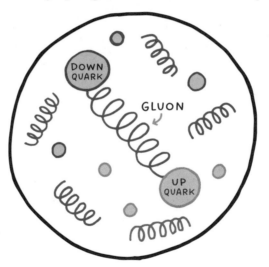

The other thing I want to make clear is that although we say quarks come in three colors—red, blue, and green—these are not actually colors. Please understand that these are just names for a property of quarks that comes in three varieties. They could have been called X, Y, Z; or rock, paper, scissors; or Huey, Dewey, and Louie instead, and been just as accurate. Red, blue, and green are just the words to describe the three types. Don't think that if you could ever see a quark directly, it would have a color.

Besides quarks, we are aware of only a handful of other elementary particles. Electrons are the simplest type of a particle, called *leptons*. Like quarks, which come in three "generations" (Up/Down, Strange/Charm, Top/Bottom), there are three generations of leptons—the electron, the *muon*, and the *tauon*. And each of these three has a partner called a *neutrino*. As far as we know, leptons and neutrinos are not made up of smaller particles.

The names and specifics of leptons and neutrinos aren't important. I list them just to give you an idea that there really aren't that many fundamental particles and that they are organized.

These twelve particles—six quarks, three leptons, and three neutrinos—form the basis of what is called the *Standard Model*. Here is a diagram:

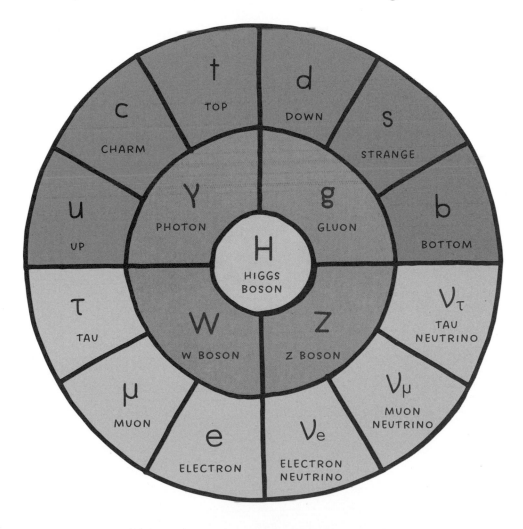

You'll notice a few particles we haven't mentioned yet. You may recall from the first chapter that we know of only four forces. In quantum mechanics, forces are transmitted by particles. The electromagnetic force is transmitted by *photons*, which are particles of light. The strong force, which binds quarks together, is transmitted by *gluons*. Similarly, the Z boson and W boson also transmit force— the *weak force* in this case. Finally, the Higgs particle is what gives all the particles mass. It was first predicted to exist in 1964 but wasn't actually found in experiments until 2013.

The Standard Model describes everything we know about the tiniest elements of the universe. However, the three generations of quarks and leptons make it seem like there's something deeper that may underlie it all. We're going to set that aside for now and return to it in our chapter The Big Bang Explained with Chocolate Chips.

But for now, think about how many varieties of cookies and other baked goods there are. Yet, at their core, they are made up of mostly the same ingredients, combined in different proportions, mixed in different ways, and baked for different times and temperatures. Similarly, everything you see around you is built of a small number of building blocks arranged in a dazzling number of patterns and combinations. It is both comforting and awe-inspiring.

CHAPTER FIVE

QUANTUM MECHANICS EXPLAINED WITH MILK AND COOKIES

When I was a kid, I used to dunk Oreos into a glass of milk before eating them.

As you dunk a cookie into milk, it creates waves that spread out from the point of impact. The waves are tough to see in a glass, but if you dunk in a bowl with enough speed, you can see them move outward.

For centuries, scientists distinguished between *particles*, like our Oreo cookie, and *waves*. They were two phenomena—something could act like a particle or a wave, but not both. In the early twentieth century, experiments started poking holes in that distinction, eventually leading to the realization that waves and particles were two sides of a much more complex coin. The new theory that described this was called *quantum mechanics*, and it upended our understanding of—literally—reality.

Waves can be described by several parameters. The *amplitude* is how high the wave is. The *wavelength* is the distance between the peaks, and the *speed* is how fast the wave moves. You also may be familiar with the term *frequency*, which is how many wave peaks you see in a second if you just watch one spot and let the waves move past you. The wavelength, frequency, and speed are related to one another—if you know two of them, you can figure out the third.

Waves have a lot of interesting features, including that they can *interfere* with one another. If you lay two waves on top of each other, it will result in a wave that is just the sum of the two. In the picture below, two waves have their peaks and valleys happening at the same time. If I send those two waves at the same time on top of each other, the resultant wave will have twice the amplitude. If those were sound waves, the sound would be twice as loud.

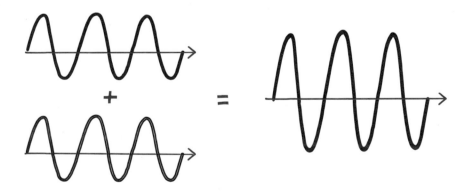

But take a look at the next picture. The waves are the same size, but now the peak of one wave is the valley of the other. If I add these together, I get the third wave shown—just a flat line! The two waves completely cancel each other out.

This is how noise-canceling headphones work. They listen for noise, and when they sense it, they create an antiwave that precisely cancels it out. But it is critical to do this precisely. If the canceling wave is slightly delayed, it will actually make the noise even louder!

This is pretty neat—and pretty useful. If you play either wave by itself through the headphones, you will hear it, and they both will sound exactly the same. But if you play them at the same time, you will hear absolutely nothing. It's as if you dunked two cookies in the milk, but with precise timing so that there are no milk waves at all. The stealth dunk.

The way that waves interact with one another to increase their strength or cancel out is called *interference*.

Particles don't create interference. If I throw two cookies at you, you're going to get hit with two cookies, regardless of how much time is in between the throws.

There's a difference between milk and cookies, between waves and particles. Waves interfere with one another. Particles don't.

Light can also create interference patterns. If you overlay light waves, they may make the light stronger or cancel it out. This is one of the observations that showed scientists that light is a wave.

That view started running into problems at the dawn of the twentieth century.

HERE'S A SELFIE
I TOOK OF ME
EATING A COOKIE:

I took this on my phone. The camera built into a phone works by focusing the light onto a special chip inside it called a photodetector.

Certain materials generate electricity when they are hit by light. Metals are one example. The electrons in metal aren't bound that tightly to their atoms. They have some freedom to move around. This is one reason metals are good conductors of electricity—the electrons can move and generate a current.

If you hit metal with light, it can knock out electrons. This creates a net positive charge in the metal, which is what your phone looks for, to see if that particular spot has been hit by light. The detector in your phone is a very dense grid of areas, each of which can be checked separately for how much light has hit it. This allows your phone to build a picture from the incoming light. A variation of this effect is also how solar panels work.

But in the early 1900s, scientists started to notice some weird stuff about those electrons. For example, scientists expected that if you shined a brighter light on the metal, the electrons that were knocked out would have more energy. But they did not. If you wanted to increase the energy of the electrons, you needed to change the *color* of the light—not the intensity. In other words, red light would kick out electrons with a certain energy, while higher-frequency blue light would kick out electrons with a higher energy.

This and other unexplained effects finally led Albert Einstein to propose a new theory of light. Einstein suggested that light wasn't a classic wave. It was packed in small bundles, called *photons*, and each photon had a set amount of energy that determines its color. So red photons always had the same amount of energy, and blue photons had a higher amount of energy. A brighter red beam had more photons, but each photon still had the same amount of energy.

Light was not just a wave—it also acted like a particle.

Scientists also were learning that particles weren't always what they thought. When they were first discovered, electrons seemed like particles. They had mass, for example.

Yet experiments and theories were showing that electrons could also act like waves. Electrons could create interference patterns just like waves!

Milk could be cookies, and cookies could be milk.

The theory and equations of quantum mechanics were developed to explain how things could be both milk and cookies—both waves and particles.

* * *

One of the consequences of everything being both milk and cookies is that we can't say precisely where an electron (or other tiny object) is or what energy it has. The possible locations are spread out across space. This is partially related to the wave nature of objects. You can't say precisely where a wave is.

Quantum mechanics lets you calculate the *probability* of finding the particle in a particular spot or with a particular energy. It does not say exactly where the particle is or how fast it is moving.

If you repeat the same experiment over and over again, you can't know exactly where you will find the particle each time. The results of one experiment are unpredictable. But the distribution of possible measurements is very predictable, as is the average measurement over time.

If I roll two dice, I don't know what result I will get. But I do know there's a one out of thirty-six chance I'll roll a twelve. And over time, I know the average roll will be a seven.

This is why the picture of the atom we are all familiar with, showing electrons orbiting the nucleus, is very misleading. An electron is not smoothly moving through space. It is better represented as a cloud of possible locations, flitting about.

You may think that while we don't know exactly where the electron is, it has a definite location at all times. We just don't know what that location is.

If you think that, you're in good company. Einstein famously said "God does not play dice with the universe" to express his disagreement with the idea at the heart of quantum mechanics that particles don't actually have a definite position until they are measured.

However, both you and Einstein would be incorrect. Experiments performed in the decades since the dawn of quantum mechanics have all confirmed that particles don't have a precise position until they are measured. They are fundamentally in a state of randomness.

* * *

Quantum mechanics is famously hard for us to comprehend. It describes a reality that bears little resemblance to the everyday. Niels Bohr, who helped develop the foundations of quantum mechanics to explain the structure of the atom as both particles and waves, famously said, "Anyone who is not shocked by quantum theory has not understood it." And decades later, Richard Feynman, who helped develop the modern theory called quantum electrodynamics (QED), said, "I think I can safely say that nobody understands quantum mechanics."

Much ink has been spilled trying to explain what quantum mechanics "means" from a philosophical standpoint, with no consensus and little understanding. I am not going to get into that here. Please refer to the Recommended Reading at the end of the book, and you can dive into this topic more deeply if you wish.

However, I would like to emphasize this: *Quantum mechanics is the most accurate and successful theory humans have ever developed*. It's weird, but it's right.

For example, there is a measurement called the *muon g−2* that quantum mechanics predicts should be:

$$2.0023318418$$

In 2021, a precise experiment measured it as:

$$2.0023318462$$

If your eyes are sharp, you'll see that the only difference is in the last two digits—sixty-two instead of eighteen.

This tiny, tiny difference has gotten physicists all excited. It means there is some interaction or new type of particle they are not accounting for in the theory. And this may crack open the door to some new physics.

This shows how the accuracy of quantum mechanics means that we need to look for the tiniest of discrepancies to try to find something new. Most predictions are on target. Quantum mechanics, with all its weirdness, is incredibly good at describing what we measure in the real world.

* * *

Quantum mechanics says that fundamentally nothing is certain. Everything is random. So if the motion of energy and particles is totally random, how does anything work? If the atoms that make up the chair I'm sitting on as I write this suddenly all move two feet to the left, I'll crash to the floor. But, while I may have clumsily missed my chair as I'm sitting down, it's never spontaneously shifted position. Why don't we experience this randomness in our day-to-day lives?

To figure that out, let's toss some cookies.

Here's a frosted sugar cookie:

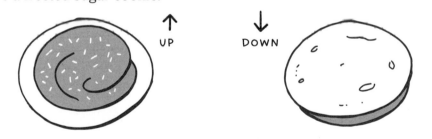

The frosting on this cookie is only on one side—a bit easier to eat that way.

If I toss this cookie into the air, half the time it will land frosting side up, and half the time the dreaded frosting side down.

If I toss two cookies, there are four ways they can land:

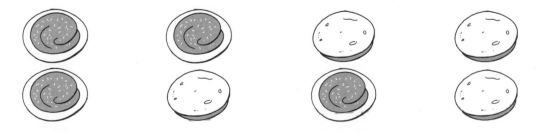

In half of these tosses, the result is one cookie up and one cookie down. Half the time it's either both up or both down.

What about if I flip ten cookies? Now it's only a 0.2 percent chance (one in five hundred) to have all up or all down. Most of the results will be pretty close to half up or half down. 85 percent of the time, there will be four, five, or six cookies with frosting side up.

Since there's a 0.2 percent chance that all ten will be either frosting up or frosting down, there's a 99.8 percent chance that there will be between one and nine frosting up. Here's a picture of a number line, showing the number of cookies with frosting side up. The brown area is the 99 percent zone. There's a 99 percent chance that if I flip ten cookies, the number that are frosting up are in that zone.

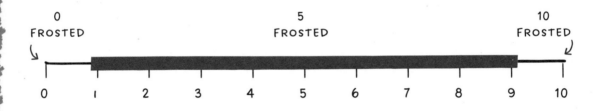

Next let's toss a hundred cookies. Now the 99 percent zone is roughly from thirty-five to sixty-five cookies frosting side up. There's a 99 percent chance a random toss of one hundred cookies will be in this zone.

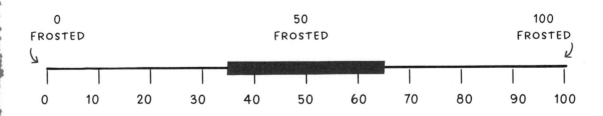

Here are the 99 percent zones for a thousand cookies, a million cookies, and a billion cookies:

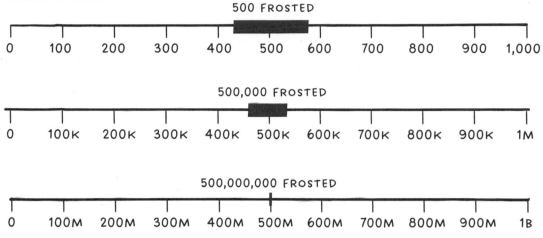

You can barely see the 99 percent zone for a billion cookies. You can be quite confident that if you flip a billion cookies the number that land frosted side up will be in that tiny brown zone.

In the introduction, I talked about how the number of atoms in a cookie is about the same as the number of stars in the universe—10^{24}. That's a one with twenty-four zeros after it. In comparison, a billion is a one with nine zeros—way smaller.

You can probably see where this is going. Our 99 percent zone is now too narrow for me to show on this page. In fact, it is narrower than the width of an atom—a trillionth of the length.

Here's another way to think about it. If I blew up the line large enough so that the 99 percent zone was the width of these pages, the line would stretch from here to the moon.

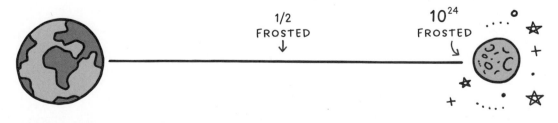

So even though all the tiniest particles behave randomly, we know with great certainty how all the atoms in a cookie will act as a whole. Is it possible all the atoms jump to the right at the same time? Yes. But the chances of it happening are so unbelievably small that you will most likely have to wait through multiple lifetimes of the universe for it to occur.

This is how we get from the weird world of quantum mechanics to our everyday world. *Lots of things acting randomly are actually very predictable.* If you throw cookies at a friend, they won't pass through them and appear on the other side. You will, quite predictably, end up with fewer cookies to eat and an annoyed friend.

CHAPTER SIX

EVOLUTION EXPLAINED WITH BUTTER AND A BAKING COMPETITION

Butter brings a lovely creamy, chewy texture to our cookie, and it is made, of course, from milk. Milk production is one of the key features that identifies an animal as a mammal—more formally, the scientific class Mammalia. Humans naturally group things that appear to be similar. We intuit that dogs and wolves are closely related, as are cats and lions. We can tell that we are closer to deer than we are to insects, and closer to insects than we are to trees. Despite the complexity and variety of life, we sense an organization to it.

Whenever we see structure and patterns like this, whether in subatomic particles, plants, or animals, it implies that it arises from some underlying mechanism. For atoms, that organizing pattern is the periodic table (Chapter Three). For subatomic particles, it is the Standard Model (Chapter Four). For living creatures, the organizing pattern is *evolution*. The synthesis of this underlying mechanism came in 1859 with the publication of Charles Darwin's *On the Origin of Species*. In that book, he lays out his theory of *evolution by natural selection*.

The core idea of evolution is that there is variation in living things within the same species, and that variation drives the characteristics of the next generation. Some deer may be faster than others, for example, or require less food. These variations change the likelihood that the deer will reproduce, creating more close copies of itself. So, over time, the population of deer will be better and better suited to the environment.

When describing his theory, Darwin said: "If it could be demonstrated that any complex organ existed, which could not possibly have been formed by numerous, successive, slight modifications, my theory would absolutely break down." Critics immediately rose to poke holes in natural selection. One of the first challenges was about mammals and milk production. This is an incredibly complex system. How could it have developed?

Darwin tackled this objection in the sixth edition of *Origins*, dedicating an entire chapter to milk. He noted that in seahorses, the eggs and young are raised in a brood pouch. Perhaps the ancestors of mammals also had pouches, and glands produced nutritious fluids for the young that were raised there.

While Darwin was not quite right with his proposal, he was close. We now have a pretty good idea of how milk production developed in mammals, and a clear step-by-step progression can be traced from early dinosaurs. The production of milk, like other systems famously put forward to disprove evolution, like the evolution of eyes, has been thoroughly explained.

For a more detailed explanation of what evolution is and how it works, let's enter a cookie-baking competition. Seems reasonable, right? Delicious, at least.

We really, really want to win, and we've got our mom's tried-and-true cookie recipe. But, with all due respect to Mom, how do we know this is the best recipe? What if we use 1 cup of sugar instead of ¾? Or a ½ cup? Or bake at 400°F, or for only five minutes?

One way—really the only way—is to try different amounts of ingredients, or different cooking times or temperatures, and taste the result. If it tastes better, try again, but move further in that direction. If I bake for nine minutes instead of ten minutes and the cookies taste better, I'll bake another batch for eight minutes and see how that works.

We can change a lot of different things each time we bake, but let's say we just decide to play with time and temperature. We bake different batches of cookies with a bunch of different temperatures, and for different times, and make a chart—maybe something like this:

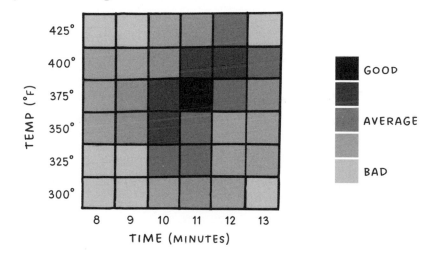

If the cookie is great, we color the square **dark blue**. If it's average, we color the square **medium blue**, and if it's bad, we go with light blue.

Hey! We found that baking the cookies for eleven minutes at 375°F is actually better than ten minutes! Now we have more confidence that we're making the best cookie. We like a nice crisp crunch to our cookie, and that combination gets us there.

The analogy of improving our cookie recipe has the ingredients (pun intended) you need for evolution to happen:

1. Instructions for how to make something.
2. A way for those instructions to change.
3. A measurement to tell how good the something is.

In this case, the instructions are our recipe, the change comes from us trying different things, the measurement is us tasting the cookie.

In the natural world, the instructions are DNA—the genetic code inside every cell. The way for those instructions to change is through *mutation*—a fancy term for copying mistakes, and *combination*—mixing up the DNA from two parents. And the measurement is provided by the environment the creature lives in—how good is it at making more of itself, of reproducing?

We go to the competition with our new eleven-minute cookies, and sadly we do not win. What happened? We scientifically tested our cookies to make sure they were the best!

When we try the winning cookies, they don't have that nice crunch we like—they are pretty gooey. It turns out our definition of "best"—cookies with a nice crunch and snap—was not the same definition of "best" the judges had. They like their chocolate chip cookies chewy.

If we want to win the competition, we'll need to change our recipe. To make a chewy cookie, we'll probably need to bake for a shorter amount of time or at a lower temperature—or maybe both.

A similar thing happens in nature. There is no objectively "best" plant or animal. It is best in a certain environment. But environments change, and what was best before may no longer be best. These changes can come from environmental changes, like global warming or ice ages, or animals or plants trying to spread into new regions, or because the other critters are changing. All these change the definition of "best."

That's why it's a mistake to say that evolution has a goal, like complexity always increasing, or toward more intelligent animals. The only thing driving evolution is what works in the environment as it changes.

There's a weakness in our cookie-contest analogy. When we set out to make the best cookies for the competition, we deliberately tried different combinations of time and temperature. But there's no intentionality in natural evolution.

Let's change our analogy in two ways.

First, let's go with a *careless cook*. Yeah, there's a recipe, but we don't always follow it exactly. It says bake for ten minutes, but sometimes we lose track of time and bake for nine minutes or eleven minutes. It says to use ¾ cup of sugar, but sometimes we're over and sometimes we're under.

Honestly, this is hardly a stretch for the way I cook.

We also make a lot of cookies, and they all vary in different ways. On average, they match the recipe, but each cookie can be a little different. Let's say that although we're careless, we still keep track of the time and temperature for each cookie and put them into the boxes on our chart from earlier:

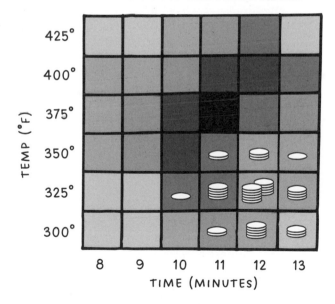

In this case, our recipe calls for cookies baked for twelve minutes at 325°F, so most of them are in that box. But we're careless, so they spread out a little. It's like a "cloud" of cookies.

In evolution, this is called a *population*. A group that is similar, but where not every individual is the same.

If we take these cookies to the competition, we can keep track of which ones people like the most. Next time, we'll make more of those, and fewer of the ones that people don't like.

We see that the cookies in the medium-blue boxes are the most popular. So next time we go to the competition, we'll try to bake more of the cookies in those boxes:

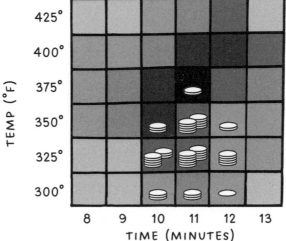

We're still careless, so our cookies are in a cloud. But we're getting closer to the most popular cookies.

But why not just make cookies in one box, whichever was most popular last time? Well, the cookies that vary from our "core" recipe are like scouts that we send out to test the other boxes. They are the explorers that allow the population to improve relative to the environment. If you have no diversity in your population, you'll never know if the box you're in is good or bad. And if the environment is constantly changing—like all real-world environments really are—you run the risk of the box you're in suddenly being really bad, indeed. And then maybe you don't get invited to the next competition. Or go extinct, if you're a plant or animal.

While in the kitchen we keep track of how the cookies are made, and which ones get eaten the most; in the natural world, this diversity and selection happens, well, naturally. Each "cookie"—each plant or animal—gets a copy of the instructions for how it's made through its DNA, but it can vary from the DNA in its parents, both from copying errors and because, for many plants and animals, the DNA deck gets shuffled when parents reproduce.

This naturally creates the cloud of diversity in the population, which will hopefully evolve into organisms better suited to the environment as it changes.

* * *

In the same way that physics ranges from studying the very large to the very small, from galaxies to quarks, biology also studies life from ecology of the entire planet to the tiny molecules that make up the machinery of the cell.

But regardless of the scale, evolution is the overarching idea that underlies all this research. It is the unifying idea across biology, and the explanatory power of evolution has been demonstrated over and over.

Yet, unlike the other scientific concepts we discuss in this book, evolution has perhaps been the most controversial, starting with its introduction in the 1850s and the challenge posed to how milk production evolved. Perhaps this is due to its challenge to the idea that humans are separate and above the rest of life on earth. When evidence started to show that the earth was not the center of the universe, that we orbited the sun, there was a similar reaction. I suppose we like to believe that everything is about us and that we occupy a privileged position.

There are some recurring arguments against evolution, which I would like to touch on briefly. One general argument against evolution is that we don't explicitly see it happen, so it is a proposition based on faith. Examples of this are:

* Yes, we've seen dogs turn into different types of dogs, but we haven't seen dogs turn into a different species.

* The fossil record doesn't show all the steps—there are gaps.

The first one is false, although a common misconception. We have seen new species develop and branch off old species. It is typically observed in creatures that have faster life cycles, just because the time for a new species to evolve is shorter. But we've seen new species created in real time in bacteria, plants, insects, birds, and more.

The fossil record is incomplete because of the nature of fossil formation. Fossils only form under specific conditions, and the plants and animals need to be in the right place at the right time to be preserved (or perhaps the wrong place from the animals' perspective!). We actually have very good fossil coverage for our ancestors the synapsids, for example.

Another is the complexity argument—systems we see in living beings are too complex to have arisen by "random chance." The argument that milk production in mammals is too complex to have developed by evolution is an early example. The development of the eye is another that is typically raised.

These objections take time and study to counter. Darwin put forward an initial hypothesis about milk production, but it took a hundred years before we developed a much better understanding of how it evolved. This is just one small slice of work that has been done on the development of systems in our body. Equally detailed work has been done on eye formation, the circulatory system, the brain, roots and stems in plants, and most other systems that form living things.

Also, I have seen the analogy that evolution is like putting watch parts into a bag and shaking them up and having a watch come out. Hopefully you've seen enough to realize that is most definitely not what it is about. It is the slow accumulation of many, many tiny changes, with a filter to help decide which changes survive into the future. It is not very complex new things just arising from chance.

People arguing complexity frequently ignore how evolution likes to find a new purpose for something that is already in use. For example, recent studies have shown that the protein *cryptochrome*, which eyes use to sense blue light, was originally part of the alarm system in cells to warn of exposure to ultraviolet light that might damage DNA. If cells were exposed to a lot of sunlight, they started to create more cryptochrome as a response, which later got duplicated and evolved into photoreceptors that helped set circadian rhythms (keeping track of day or night), and ultimately into blue receptors in eyes.

Life is a master at recycling. Different functions are constantly getting repurposed and combined to do new things. Flour was first used to bake bread. But look at the myriad of uses it has been put to over thousands of years. Every culture has its own distinct spin on breads, cakes, cookies, and pies. But flour can also be used as a cleaner, glue, insect repellent, and skin treatment. Bakers keep finding intriguing and novel ways to combine the ingredients in this book. Similarly, the evolutionary process uncovers new uses for proteins all the time.

Of course, like every single idea in this book, scientists keep learning more about how evolution works. We are constantly modifying the details of the theory, but the core remains the same.

For example, evolution seems to proceed sometimes gradually, and sometimes rapidly. Why is that?

Another thing we are learning about: Complex genetic switches in our bodies determine which genes are active and how much protein they produce (called *epigenetics*). How do these work? Epigenetics can get passed down to future generations, so even if you have the same DNA, your epigenetics may differ.

Finally, we are learning more and more that the microbes that live in our guts and on our skin have a profound effect on how our bodies operate. There are ten times as many cells that are not us as those that have our DNA. How does that work? What coevolution happens between us and our microbiomes, and how is it affected by climate, diet, and even culture? These are still open questions.

Evolution as described by Darwin in 1859 isn't perfect. We have learned a lot more in the last 150 years. But the framework of *variation plus selection* is incredibly powerful. If we encounter life elsewhere in the universe, it is highly unlikely that it will use exactly the same DNA molecules that we do to store genetic information. It won't produce the same milk and butter as mammals here on earth. However, we can be quite certain it will have grown and developed according to natural selection.

I look forward to introducing our new alien friends to the world of baking competitions.

CHAPTER SEVEN

GENETIC ENGINEERING EXPLAINED WITH AN EGG

In the last chapter, we talked about how one of the key requirements for something to evolve is a way to record and transmit the instructions on how to build it—the recipe, so to speak. There needs to be a recipe for everything that needs to be built.

* * *

For all of us—animal, plant, bacteria, and all life—that information is encoded into the DNA at the heart of our cells. The chicken is no exception.

The egg is an important symbol in many cultures. It often symbolizes springtime, rebirth, and creation.

But from the chicken's point of view, the purpose of the egg is clear—to make more chickens.

A fertilized egg contains the instructions for making a new chicken, the machinery to follow those instructions, and the energy and raw materials to make it happen. In this chapter, we'll look at how the instructions are written. Then in the next chapter, we'll learn about the machinery.

* * *

Bad news! You've been trapped in a cookie factory and are being forced to bake. Your only hope for rescue is to get a message to the outside world. Unfortunately, you don't have anything to write with, so you're going to have to come up with some other solution. How can you send a message?

The only tools you can use are the cookies you are baking. You can bake four types:

ANIMAL CRACKERS

CHOCOLATE CHIPS (NATURALLY)

GINGERSNAPS

THUMBPRINTS

You are constantly watched, so you can't change the cookies at all. However, you can arrange how they are packaged—you can put them in any order you want. Maybe you can use that somehow? Create some type of code?

Your message is going to be composed of words. There are many, many words you might want to express in your message, so it's going to be very difficult to encode words directly.

But words are made up of letters—and even though there are thousands of words in the English language, they are all composed of just twenty-six letters. Can we use the cookies to represent letters?

If we just use one cookie, we could represent only four letters.

If we use a group of two cookies together, we increase the number of combinations from four to sixteen (four times four):

...and so on. You get the idea.

That's close to twenty-six, but we still don't have enough to represent every letter. To do that, we need groups of three cookies. That gives us four times four times four combinations—sixty-four in total. That's plenty to represent the twenty-six letters, plus we can have combinations for the digits from zero to nine, and even spaces, commas, and periods. Now we're getting somewhere!

You decide on a representation for each letter—maybe something like:

= "A"

= "B"

= "C"

...and so on, and make a nice little chart.

2ND COOKIE

1ST COOKIE / **3RD COOKIE**

A	B	C	D
E	F	G	H
I	J	K	L
M	N	O	P
Q	R	S	T
U	V	W	X
Y	Z	0	1
2	3	4	5
6	7	8	9
SPACE	.	,	!
@	#	$	%
&	*	(	)
-	=	+	/
<	>	:	;
"	'	[	}
{	}			?

You carefully arrange the cookies in the package three cookies at a time to spell out your message. If you want to start with "HELP!" the cookies would be:

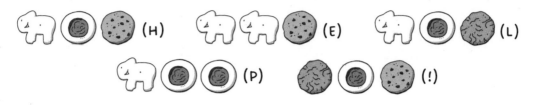

After a lot of baking, you manage to put together your full message:

HELP! I'M BEING HELD CAPTIVE IN A COOKIE FACTORY!

Someone notices the order of the cookies is very odd and posts pictures on the internet. It goes viral, and millions of people work on it until the code is broken, and you are rescued.

A happy ending to our tale. Well done!

* * *

Our cookie-factory story, while also being a very dramatic tale and the prime candidate for a spin-off Netflix series from this book, also will help us understand DNA and the machinery that lies at the heart of an egg, and at the center of every cell in our bodies.

Remember that one of the requirements for evolution is a way for the pattern that makes a living thing to be reproduced. There needs to be a recipe with instructions for everything that needs to be built.

That message is analogous to the organism—let's say the chicken in this case.

ESCAPE MESSAGE = THE CHICKEN

When we were trying to escape the cookie factory, our message was composed of *words*. In a language, a word represents a single concept—an object like "cookie," an action like "bake," or descriptor like "yummy." In living things, words are equivalent to *proteins*. Proteins perform all the functions required to keep your body running. They help chemical reactions occur, including creating and using energy; sense the environment; act as gatekeepers in cells to help stop invaders; contract to power muscles; provide structure to cells; send signals; and move molecules from place to place. About twenty thousand proteins make up the human body. But each protein almost always performs just a single purpose. Hemoglobin, for example, holds oxygen molecules and transports them to your cells. Insulin regulates your metabolism, telling your body when to absorb sugars.

In the same way you can combine words in a vast number of ways, from a novel to a birthday card to a cookie recipe, you can combine and organize proteins in a vast number of ways, to create all the types of living beings we see on earth, from bacteria to ants to whales.

WORDS = PROTEINS

Just as words are composed of letters, proteins are composed of building blocks called *amino acids*. Think of them as beads on a string, or—to build on our analogy—letters in a word.

Our bodies use twenty amino acids, a very similar number to the twenty-six letters in our alphabet. Like letters, those twenty amino acids combine in different ways to form the twenty thousand proteins we use. Coincidentally, most of us have a vocabulary of around twenty thousand words—so we use twenty-six letters for twenty thousand words, and our bodies use twenty amino acids to create twenty thousand proteins.

LETTERS = AMINO ACIDS

Letters is as far down as our language goes. But in cells, the "letters"—the amino acids—are themselves encoded.

DNA

In cells, DNA is ultimately where our "recipe" is stored. DNA is composed of just four different molecules linked together in a giant chain. The pattern of those four molecules is ultimately the recipe for life.

The four molecules are called adenine, cytosine, guanine, and thymine, but they are commonly referred to by just their starting letters—A, C, G, and T.

I like to call them animal cracker, chocolate chip, gingersnap, and thumbprint.

When we were stuck in the cookie factory, we needed to use groups of three cookies to represent all the letters of the alphabet, and also some special characters like punctuation. Same thing with DNA. To code for twenty amino acids, the four "nucleic acids," as they are known, are read by the cell's machinery in groups of three.

Remember the table we made to figure out our cookie code? DNA works the same way. Here is our cookie table from earlier, and the real "genetic code" that is used in all living things:

1ST COOKIE	2ND COOKIE				3RD COOKIE
	A	B	C	D	(animal cracker)
(animal cracker)	E	F	G	H	(chocolate chip)
	I	J	K	L	(gingersnap)
	M	N	O	P	(thumbprint)
	Q	R	S	T	(animal cracker)
(chocolate chip)	U	V	W	X	(chocolate chip)
	Y	Z	0	1	(gingersnap)
	2	3	4	5	(thumbprint)
	6	7	8	9	(animal cracker)
(gingersnap)	SPACE	.	,	!	(chocolate chip)
	@	#	$	%	(gingersnap)
	&	*	(	)	(thumbprint)
	-	=	+	/	(animal cracker)
(thumbprint)	<	>	:	;	(chocolate chip)
	"	'	[	}	(gingersnap)
	{	}	|	?	(thumbprint)

1ST NUCLEOTIDE	2ND NUCLEOTIDE				3RD NUCLEOTIDE
	A	**C**	**G**	**T**	
A	LYS	THR	ARG	ILE	A
	ASN	THR	SER	ILE	C
	LYS	THR	ARG	START	G
	ASN	THR	SER	ILE	T
C	GLN	PRO	ARG	LEU	A
	HLS	PRO	ARG	LEU	C
	GLN	PRO	ARG	LEU	G
	HLS	PRO	ARG	LEU	T
G	GLU	ALA	GLY	VAL	A
	ASP	ALA	GLY	VAL	C
	GLU	ALA	GLY	VAL	G
	ASP	ALA	GLY	VAL	T
T	STOP	SER	STOP	LEU	A
	TYR	SER	CYS	PHE	C
	STOP	SER	TRP	LEU	G
	TYR	SER	CYS	PHE	T

If you understand the cookie chart, you should be able to understand the nucleotide chart—they're the same thing. The full chemical names of the amino acids are abbreviated—"LYS" is called "lysine" and "THR" is "threonine," for example—but the names aren't important. The key idea is that three letters are coding for a single amino acid.

If you look closely, you'll see that the code ATG represents START. This code shows where each protein starts on the DNA. There are a few codes that represent STOP—TAA is one of them. The code for each protein lies between a START and a STOP. These are analogous to the punctuation marks in our cookie code—spaces and periods.

 ➡ A, C, G, T

LETTERS, PUNCTUATION ➡ AMINO ACIDS, STARTS, AND STOPS

WORDS ➡ PROTEINS

MESSAGE ➡ ORGANISM

The human genetic code has about three billion nucleotides in our DNA. The bacteria *E. coli*, in contrast, has only about five million. But we're far from having the most DNA. The Norway spruce has twenty billion, but the crown goes to a rare Japanese flower, Paris japonica, which has a whopping 150 billion—fifty times as many as humans.

Our three billion nucleotides means that we have one billion "letters" in our code, since each triplet corresponds to a letter/amino acid. In comparison, the longest novels are around four million letters.

And a chocolate chip cookie recipe has 668. So longer is not always tastier.

CHAPTER EIGHT

EMBRYONIC DEVELOPMENT EXPLAINED WITH COOKIE DECORATING

While the instructions are one key part of the egg, another is the machinery to read and perform those instructions. It seems almost miraculous that a baby can grow from a single cell and that all the patterns that form our body are put into the right spots.

Fortunately, over the last few decades, we've learned a tremendous amount about how these patterns are formed. It's best explained through the use of...a cookie.

I want to create a cookie with a nice design on it, done in frosting. But I'm super lazy, and I don't want to put the frosting on myself. Instead, I've invented a tiny little machine that can put frosting right in its area.

I've got a blank cookie, and I've scattered thousands of these little frosting machines across the face of the cookie, like pixels on a television screen. I can give each of these little dot machines instructions on when to make frosting and when not to.

But because I'm lazy, every single machine has to have the same instructions. I only want to write one set of instructions, not thousands!

How can I set this up to make an intricate pattern?

Before we solve this problem, I will tip my hand and confirm what you probably already suspect. This self-frosting-cookie problem is analogous to how an embryo develops. Embryos start out as one cell, but quickly they duplicate into a mass of thousands of cells. Somehow the cells need to decide if they are going to make a bone or a muscle or skin or an eye. But every cell has the same instructions—they all share the same DNA.

Looking at the self-frosting cookie will help explain how this works. We'll then talk about some of the differences between our simple cookie model and the real world.

We are all self-frosting cookies.

OUR SELF-FROSTING TOOL KIT

Here is our blank cookie:

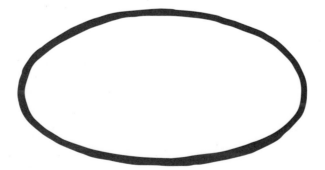

To create our designs, we need a few features.

First, we need some type of signal that is strong on one side of the cookie and gets weaker as you move across. Something like this:

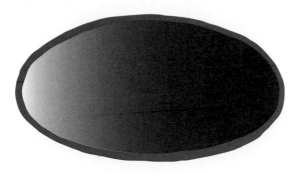

It could be a chemical, or an electric or magnetic force, or light, or anything really. But whatever it is, our little frosting machines have sensors that can tell the strength of the signal. Then our machines need to have instructions as to whether they are "on" or "off," based on whether the signal is above or below a certain value.

For example, let's say our signal goes from zero to one hundred. We could have a rule for the machine to turn on when it sees seventy or higher. Then the machines in the black area will turn on.

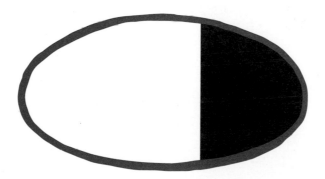

TURN ON IF YOU SEE SEVENTY OR HIGHER

If the rule is turn on when it's forty or higher, it will look like this:

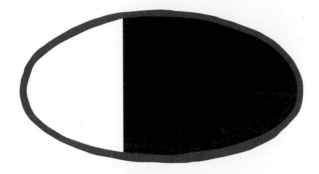

TURN ON IF YOU SEE FORTY OR HIGHER

We're getting there—these patterns look like we dipped our cookie into frosting. But we can probably do better.

WAVE IT UP

The next piece of our tool kit is another type of signal—and the only other one we will need. This one is in the form of a wave. It looks like this:

This wave signal, and the earlier one that goes evenly from high to low, can both be generated by a wide variety of natural processes. So they are good to use as fundamentals for our system.

We can use our "on/off" switch with the wave signal. This gives us stripes! Now we're getting somewhere! This looks like a proper frosted cookie.

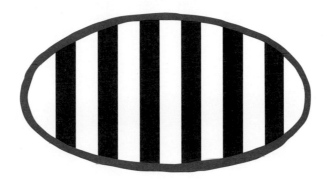

STRIPE SIGNAL, TURN ON WHEN FIFTY OR HIGHER

We don't have to generate the signals from left to right. We can also create them from top to bottom. That lets us do something like this:

One final piece for our tool kit. We know how to make blocks of color and stripes. We used them to turn our switches on, but what if we could use them to turn a switch that was on back to off?

In this diagram, I've got two patterns. The black one turns machines on, but the brown pattern turns machines off. The "off" patterns are generated using the same rules—so blocks and stripes.

Combining those two leads to the third diagram—a stripe that is off center.

START WITH THE BLACK PATTERN
(MACHINES THAT ARE HIGHER THAN FORTY TURN ON),
AND SUBTRACT THE BROWN PATTERN
(MACHINES THAT ARE HIGHER THAN EIGHTY TURN OFF).

You can see that by varying how big the different blocks are, we can put a stripe anywhere with any thickness.

And now, with our full tool kit, we can do lots of fun stuff! We can take a horizontal block and subtract a vertical block and make an area in the corner:

Or we can take vertical stripes and subtract out horizontal stripes to get a checkerboard pattern:

Or we can start with stripes but subtract out three separate blocks to just leave us with three small patches:

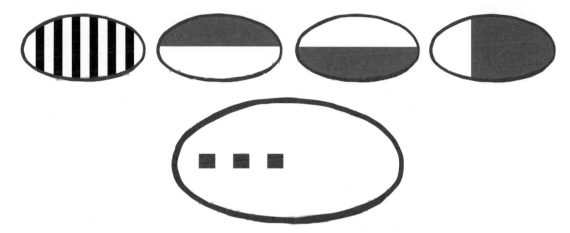

Using these tools, you can create a dizzying array of patterns, while still following the restrictions set by our lazy cookie decorator.

These procedures are essentially the tools that an embryo uses when it is developing.

The "signals" in this case are protein concentrations. They are high near one end of the embryo and drop off as you move across. It is easy to make this by just having a "pump" at one end of the embryo that puts out the protein. Then it is high near the pump and naturally drops off as you move farther away.

Very early in its development, the embryo develops two "axes"—an east-west gradient, and a north-south gradient. This enables each cell to, in essence, figure out its coordinates in the embryo, much like using latitude and longitude on a globe.

These signal proteins activate genetic *switches* on the DNA, which are similar to the switches in our cookie-frosting machines. How do these switches work?

Earlier we talked about how DNA gives the sequences of amino acids to make proteins. This is only partially true. Only 1 percent of DNA actually encodes proteins. For a long time, we used to think the other 99 percent was "junk" DNA that didn't do anything. Maybe it was just a leftover from evolution. However, in the last few decades, we have come to realize that this so-called "junk" is anything but. Much of it is for these genetic switches that either start or stop production of certain proteins.

Signal proteins look to latch onto a specific sequence of DNA. When they find it, they clamp on. Sometimes, depending on the shape of the protein, this attracts attention from the "machine" that reads the DNA (called RNA polymerase). These are called *promoters*. But sometimes the signal protein will instead *prevent* the RNA polymerase from attaching to the DNA and reading it. These are called *repressors*.

Through this combination of promoters and repressors, cells in some areas will activate certain sections of DNA, and others will not. This is completely analogous to how the cookie-decorating patterns are generated.

For example, let's look again at this combination from earlier:

Remember, we said the final area (with the blue) happens by starting with the black area and subtracting the brown area. What's happening on the DNA to make this happen?

Here's a gene that when it's read, it produces BLUE:

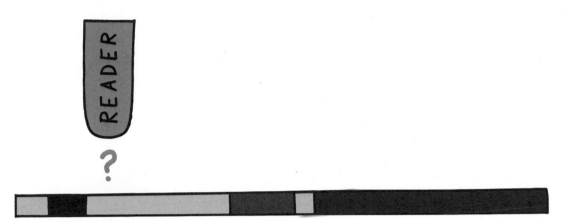

There's no helper protein here, so the reader doesn't produce blue that much, if at all. It needs help to clamp onto the DNA properly.

Let's say that a cell that is in the BLACK part of the embryo shown on the last page has the machinery turned on to produce a protein called BLACK—and it floats around the cell. When it floats by the black section of the DNA strand above, it wants to latch on, since it matches that part.

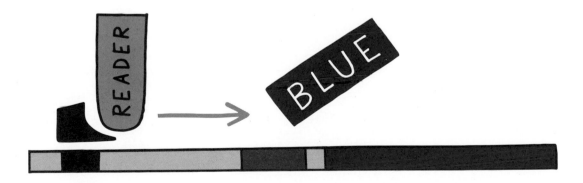

When BLACK latches on, that creates a friendly ramp for READER to start at. It happily slides down the DNA and produces the BLUE protein.

But if the cell is also in the brown portion, another part of the DNA produces BROWN.

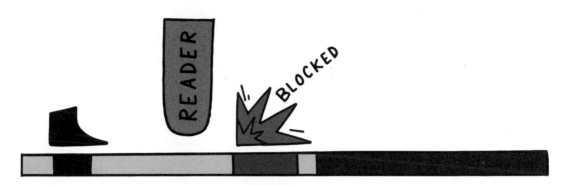

BROWN latches onto the BROWN section. It is spiky and generally unfriendly, and the READER is stopped when it hits it. BLUE is no longer produced.

The combination of these factors results in BLUE being produced only in cells that are producing BLACK but aren't producing BROWN. BLACK is a promotor, and BROWN is a repressor.

One thing we haven't talked about is what BLUE does. That's the neat part— BLUE could perform a function in the cell, like moving sugar around, but BLUE could also be a signal—either as a promoter or a repressor for other proteins, or even as a promoter for some or a repressor for others.

In this way, the entire organism is built up.

This example from earlier is actually similar to a real pattern that you see during the development of many animals.

Those purple squares are where limbs will grow. The "purple" in this case is a special protein called "distal-less," or Dlx for short. It got that name because when researchers interrupted it, leg development stopped.

When Dlx is present in an area, a limb will grow out of it. It's a switch that turns on the entire "make a leg" program in cells in that area.

Similarly, a protein called PAX6 causes eyes to be formed. Wherever PAX6 signals, the eye "program" is run. Researchers have done experiments on fruit flies where, for example, they put PAX6 on the embryo where the wings develop, and sure enough there was an eye on the wing.

As it became easier and easier to sequence DNA, researchers made a startling discovery about genes like Dlx and PAX6. They are in all animals, are almost exactly the same proteins, and do almost exactly the same thing.

If you put a mouse PAX6 into a fly, it still develops an eye. A fly eye, in case you were wondering.

This means that the concept of an "eye" must be incredibly ancient, and the meaning of "eye" has been preserved through just about every animal species. Evolutionists assumed that eyes as different as the fly's compound eye and our eye evolved separately or diverged so long ago that the mechanisms for their construction would be radically different by now. But they're not. That master "make an eye here" switch has been preserved for hundreds of millions of years. Even simple worms that just have some light-sensitive patches use PAX6 to trigger their formation. And going back to limbs, snakes contain the Dlx signal that produces them. But it is never produced in snakes because of repressors.

This modular structure to animal growth—put legs here in one creature, there in another, eyes over here, make the midsection longer or shorter—gives evolution tremendous power in developing new body forms. It makes building an animal akin to a set of Lego bricks that snap together in an endless variety of ways. But they always use the same core genetic tool kit.

We are all, in essence, cookies with incredibly complex decorations.

CHAPTER NINE

UNCERTAINTY EXPLAINED WITH 3/4 CUP OF PACKED BROWN SUGAR

At 12:30 PM on January 15, 1919, the residents of Boston felt the ground start to rumble. A loud roaring noise quickly followed as a tank containing millions of liters of molasses collapsed, releasing the sticky substance.

A giant brown wave of molasses rushed down the streets, trapping people and animals in its syrupy embrace. The *Boston Post* reported:

"Molasses, waist deep, covered the street and swirled and bubbled about the wreckage...Here and there struggled a form—whether it was animal or human being was impossible to tell. Only an upheaval, a thrashing about in the sticky mass, showed where any life was... Horses died like so many flies on sticky fly-paper. The more they struggled, the deeper in the mess they were ensnared. Human beings—men and women—suffered likewise."

The Great Molasses Flood, as it became known, killed twenty-one people and injured 150. The cleanup took weeks and weeks, and everything in Boston was sticky. Residents reported that for decades afterward, on hot days the smell of molasses hung over the city.

THE AFTERMATH OF THE GREAT MOLASSES FLOOD

Brown sugar is sugar that contains some molasses, and molasses is created by boiling the extract from sugarcane. For pure molasses, it is boiled and then cooled three times. Each cycle extracts more sugar and concentrates the liquid, making it darker and thicker.

Brown sugar is the result of the first boiling, when a bit of molasses is still mixed in with the sugar. However, most brown sugar that you buy in the store is created by blending pure sugar and molasses together, and is about 3.5 percent molasses. Dark brown sugar is about 6.5 percent molasses.

* * *

What caused the Great Molasses Flood? There was a big lawsuit against the tank owner—the Purity Distilling Company. (Why distilling? Molasses can be fermented to make ethanol.) It took years to be resolved. Did the molasses itself explode? Was it a bomb by Bolsheviks or anarchists? Or a failure in the tank itself?

Ultimately, the court found that the tank was to blame. It had not been properly inspected and was poorly designed. It had been built only four years prior and had been filled to capacity only eight times since then.

Most important, the day before had been very cold, but temperatures warmed up rapidly, going from 2°F to over 40°F on the day of the flood. Plus, the tank had just been filled the day before. The rapid warming caused the molasses to expand, which put pressure on the tank, resulting in the collapse.

Whenever something is designed, engineers need to deal with *uncertainty*, about the materials used for construction, the way it is assembled, or how it is used. You probably are not familiar with welding and riveting steel plates to build a storage tank. But you probably are familiar with that most famous of cooking construction projects—the gingerbread house.

If you're a gingerbread engineer (a gingerbrengineer?) designing a gingerbread storage tank, you may build some sample tanks first and figure out how thick the walls need to be to hold your molasses. But when you actually build the final tank, there will definitely be differences between that and your tests. Maybe some of the gingerbread was baked longer or the ingredient mix was slightly different. The icing you use to hold the pieces together may be stickier—or not as sticky—as your test icing. You can imagine all the ways a gingerbread building will be different each time you construct it.

So what is a gingerbrengineer to do? How do you compensate for all this variation?

You build in a *safety margin*. If you calculate that the strength for the amount of molasses is a certain value, you design a tank that can hold double that. Then, if there are variations in the materials or construction, you can be fairly certain the tank will still be okay. It's not guaranteed, of course, but you can be much more confident.

The tricky part is deciding what safety factor to use. In our example, we used double—2x the strength. But making things stronger will make them more costly. So if you are a company that makes tanks, you might consider reducing the safety margin. If 2x is good, then 1.95x is almost certainly okay as well, and you will probably reduce the cost of your tanks and be more competitive.

Even though tank failures can be catastrophic, they are also rare and require a combination of factors, like the rapid rise in temperature just before the Great Molasses Flood and a just-filled tank. The free market doesn't handle situations like that very well. Companies will be tempted to continue to shave from their safety margins to compete with one another, until disaster strikes.

This is a great example of where government regulation can help. If the government mandates a specific safety factor, that creates a level playing field for all businesses and a defined level of safety for the public. Self-regulation does not work for this sort of issue. An external group that has the power to inspect and enforce certain requirements is much more effective.

★ ★ ★

Another possible source of error, both in measuring the ingredients for our gingerbread house and testing its strength is the measurement itself. When you measure out ¾ cup of brown sugar, how accurate is it really?

There are several possible sources of error: You could not fill up your cup exactly level—perhaps the brown sugar is a little below the surface or is heaping up above. Or maybe it isn't packed down as much as the recipe writer meant it to be.

However, it's also possible that your measuring cup doesn't hold precisely three-quarters of a cup. How do you know how accurate your measuring cups and spoons really are?

Devices used for measurement need to be tested against a known good standard. This is called *calibration*. Some equipment can be adjusted to make the readings match the standard. In other cases, like with measuring cups, they cannot be changed. If the error is small enough, they might still be okay to use, depending on the purpose. Or they might need to be taken out of service.

Calibration and testing extend well beyond the world of science. Measures used for commercial reasons also need to be tested. Government officials typically test supermarket scales and gas pumps annually to ensure they are accurate within accepted standards.

In fact, making sure that commerce is fair and equitable was the motivation for the invention of measurement systems. One of the oldest was developed in Babylonia five thousand years ago, where the standard of weight was the *shekel*. Sixty shekels made a *mina*, and sixty minas made a *talent*. Standard weights were distributed to help ensure fair trade.

When an inspector calibrates a scale at the supermarket, they bring along a set of standard weights. But how do they know those weights are accurate? The weights themselves need to go to a lab to be checked and calibrated. But how do we know the lab's equipment is correct? It also needs to be calibrated.

This is called the *calibration chain*, and it ensures that everything is controlled and accurate. But what is at the top of the chain? What is everything ultimately compared with?

One of the oldest units—the cubit—was the distance from the elbow to the tip of the middle finger. As can be imagined, this will vary quite a bit from person to person. Sometimes it was based on the ruler's arm, but again, that could vary, even over someone's lifetime. In Egypt and other ancient civilizations, a standard rod was created and stored safely. Then other rods of matching length were created based on the master and distributed throughout the region.

A unit of weight is hard to base on the human body, but fortunately grain is fairly regular in size. The *grain* was often a unit of weight and is, in fact, still used today.

Having reference samples like standard rods does not fulfill the requirements of modern measurement. We need something that doesn't need to be created by humans, that can be found in nature, and is very stable. The International Bureau of Weights and Measures, based in Paris, is responsible for developing these standards.

For example, the second is officially defined as:

THE DURATION OF 9,192,631,770 PERIODS OF THE RADIATION CORRESPONDING TO THE TRANSITION BETWEEN THE TWO HYPERFINE LEVELS OF THE GROUND STATE OF THE CESIUM-133 ATOM.

I won't go into exactly what this means, but it has to do with electrons moving between different energy levels in atoms. This is something labs everywhere can measure. Alien life-forms could also understand this and determine the length of our second. Cesium is cesium, everywhere in the universe.

Once we have a second defined, we can build on that. The meter, the base unit of length, is defined by the speed of light, which is also constant:

THE LENGTH OF THE PATH TRAVELED BY LIGHT IN A VACUUM DURING A TIME INTERVAL OF 1/299,792,458 OF A SECOND.

Improving these systems of measurement is something that is ongoing. The kilogram, the unit of mass, originally was defined as the mass of a liter of water. However, this wasn't all that accurate—water can have different substances dissolved in it, and the density can vary quite a bit based on temperatures and pressures.

In 1889, a cylinder was created out of platinum-iridium and became the standard kilogram for the globe. However, as you can imagine, over time it could lose atoms, and a better standard needed to be found. It wasn't until just a few years ago, in 2019, that an official replacement for the kilogram was found, based on *Planck's constant*, which is a key value from quantum mechanics and determines how much energy a photon has.

Now all measurements are based on universal constants. That ¾ cup of brown sugar can ultimately be traced back to cesium radiation, photon energy, and the speed of light. Not bad for something in your kitchen drawer.

CHAPTER TEN

THERMODYNAMICS EXPLAINED WITH BAKING AND AN ICE CREAM SANDWICH

In 2019, a special oven was sent up to the International Space Station to be used for baking, and it was first used to—you guessed it—bake a batch of chocolate chip cookies.

We cook something to increase the temperature. Usually this causes a physical or chemical change, like melting chocolate chips or causing a souffle to set. But it can have a lot of other effects, like causing air pockets to expand to make a cake fluffier or killing bacteria.

Regardless of the goal, conventional ovens typically work the same way. Heating elements on (usually) the top of the oven raise the temperature, which raises the temperature of the food.

WHAT IS HEAT?

How does the heat get from the heating element to the cookies? And what exactly is "temperature"? There were several theories in the 1700s, but the established candidate was "caloric theory." This stated that heat was an invisible fluidlike substance called "caloric" that flowed between objects (from hot objects to cold objects, warming them up). However, in the 1800s, scientists discovered that heat is simply molecules jiggling back and forth. The faster they jiggle, the "hotter" they are.

In a typical oven, the heating elements at the top and bottom are heated up by electricity. But the heat needs to get from there to the cookies. It does so by colliding with other molecules, like trillions of tiny billiard balls on a pool table.

Initially, all the air molecules in your oven move at roughly the same speed—they have the same energy. As the element heats up, some of the molecules hit it, pick up some energy, and start moving faster. When they collide with slower-moving molecules, some energy gets transferred. And gradually, over time, the average energy of all the molecules starts to rise.

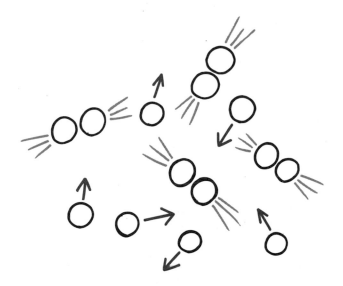

This has a number of consequences. First, it will take some time for your oven to heat up, for all these collisions to happen. Second, the temperature closer to the heating elements will be hotter than that in the center of the oven. This variation can sometimes be quite significant.

If you bake two sheets of cookies on separate racks, you'll find that the cookies on the bottom lag behind the ones on the top rack because they are farther from the heat source.

The process of heat spreading through an area by collisions is called *con-duction*. As the air molecules collide with the cookie, they transfer energy to its molecules. The air molecules lose energy in the process, cooling down. Then they hang around, waiting for hotter, higher-energy air molecules to hit them, to make them energetic again, so they once again can collide with the cookie. That, as you can imagine, takes time.

What would help is, once molecules lose their energy to the food, they moved out of the way and let higher-energy molecules take their place. An energy conveyor belt, if you will.

This type of system is called *convection* and is what is used in (not surprisingly) a convection oven. A fan moves hot air around the inside of the oven, which gets those colder molecules out of the way, back toward the heating element, while hot air takes its place.

You can see a similar effect another place in your kitchen—a pot of boiling water. Colder water is denser than hot water. So as water is heated from the bottom, the hot water rises to the top, and the cold water falls, the same way that hot air rises. That brings the cold water close to the fire and heats it up faster, continuing the cycle.

If you were to boil water from the top instead of the bottom, it would take a lot longer. I'm not sure why you'd try, but just so you know.

As you can imagine, it takes a while for energy to get transferred from the heating elements in your oven to the food in the center. It'd be great if the food could heat up directly, without requiring all that random bouncing through the air.

Well, that's how a microwave oven works. Microwaves pass through the air without interacting with it. However, they can transfer energy directly to water molecules. Water molecules in food energetically vibrate when hit by micro- waves and then bounce into the other molecules in the food, transferring the energy to them. All the conduction happens inside the food—the whole step of conducting heat from the heating elements to the food via the air is skipped.

So what happened to that special oven that was sent into space?

The cookies took longer than normal to cook, and they also didn't spread out as much. Both of these effects are mainly due to the zero-gravity environ- ment. Even in conventional ovens (non-convection), a little bit of air still moves inside the oven because of temperature differences. In zero gravity, hotter air will not rise.

If you boil a pot of water in space, there also is no convection in the water. All the steam bubbles would stay at the bottom of the pot. Bubbles rise in a pot of water because gravity pulls the water downward more than the bubbles. As the water gets pulled down, the bubbles get displaced and move upward. If there is no gravity, nothing pulls the water down.

★ ★ ★

Let's turn our attention from heating things up to cooling them down. For me, one of the highlights of summer is enjoying an ice cream sandwich on a hot summer day. The vanilla ice cream nestled between two chocolate chip cookies is an unbeatable combination.

Ice cream goes back to at least the 1500s—well before the invention of freezers. To make ice cream, the temperature of the ingredients needs to be lowered to well below freezing to make the semisolid, creamy texture. Ice cream is typically served at around −6°F.

How do you manage this trick? How do you get ice cream lower than 32°F if you want to make it in the heat of summer and you live before the invention of freezers?

The key is to add salt to the ice. Salt water has a lower melting point than plain water. The salt particles in the water make it harder for the water molecules to stick together in a solid. So salt water needs to be colder than normal water to solidify. This also means if you start with ice, it will turn back into water at a lower temperature.

If you live in a climate with snow and ice, you probably recognize that salt is put on ice to melt it. This is the same idea.

If you want to make ice cream, you place ice around your bowl of ingredients and then mix salt in with the ice. This will reduce the temperature of the ice down to the melting point of salt water—about −6°F, our target temperature for making ice cream!

But how does the temperature drop? The laws of thermodynamics say that the total heat in a system stays the same. What temperature is going up in this system? The ice, and the ice cream ingredients are all getting colder. If I take a cookie out of the oven, still warm, and dunk it into a glass of cold milk, they will both change temperature. The cookie will cool down, and the milk will warm up. The energetic cookie molecules will collide with the slower-moving milk molecules and transfer energy over to them.

But if the cookie starts out at 150°F, and the milk at 40°F, the cookie will never get colder than 40°F, or the milk warmer than 150°F. The final temperature will be somewhere in between. We can't get our cookie temperature below 40°F— forget about −6°F.

So what's going on?

When ice is brought out into the sunshine, the surface quickly warms up to 32°F, and then it stays there while the ice melts. While the ice is melting, the temperature stays right at 32°F.

Here's a graph of what the temperature of the ice cube looks like as it's melting:

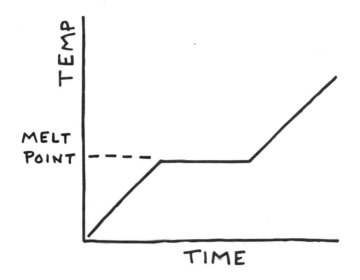

The temperature rises until it hits the melting point. While the ice cube is melting, it stays at the same temperature. Then, when it is completely changed to liquid water, the temperature will start rising again.

Remember that the temperature is a measure of how much the molecules are vibrating. It's not (necessarily) the amount of energy that is in the system.

When the ice cube is at the melting point, additional energy from the sun (or whatever is warming up the ice cubes) doesn't go into increasing the vibration of the molecules. It goes into separating the water molecules from the solid ice state into the liquid water state.

The same thing happens when water boils. The temperature stays at 212°F until all the water has converted to steam, and then the temperature of the steam can start to rise.

This, by the way, is a big reason we cook with boiling water—it's a known temperature, so you know how long to leave your pasta in the pot, for example. You don't have to constantly regulate the temperature or worry that your water is overheating. It literally can't.

This is also why you feel cooler when you sweat in the hot sun waiting for me to explain to you how to make ice cream. Sweat is at the temperature of your body. But it needs to absorb an extra bit of energy to change from a droplet on your skin to a gas—to evaporate. This heat comes from your skin, and you feel cooler. Even though the temperature of the droplet hasn't changed, energy has been drawn out of the system to evaporate the sweat drop.

When you add salt to ice, something similar is going on. The ice-salt mixture on the outside of the ice cubes is above its melting point—so it turns into water. But to do that, it needs to pull heat from somewhere—in this case the ice cream ingredients, as well as the surrounding ice that is still at 32°F. As the salty ice continues to melt, it pulls the temperature further and further down until everything reaches −6°F.

As a fun and tasty science experiment at home, you can make a basic form of ice cream with some zip-top bags. Take a small one, put in some milk, sugar, and vanilla, and seal it up. Then take a gallon-size bag, fill it with ice and rock salt, put the small bag inside, and zip up the large bag.

Shake it up for five minutes, and you'll have ice cream! If you try this, you should probably wrap up the bag in newspaper or wear gloves. The bag will get really cold really fast, thanks to the interaction of the ice and salt. Then enjoy your tasty treat while remembering that temperature is not the same as energy.

During the 1800s, scientists spent a lot of time researching energy and heat, and *thermodynamics*—literally "the power of heat." They boiled the knowledge (pun intended) down to three "laws."

The first law of thermodynamics says that energy in a closed system is conserved.

As we talked about in Chapter Two, energy can be converted between different types, but it isn't created or destroyed. This sounds odd—we generate energy all the time, right? The key phrase here is "closed system." Your body, for example, is not a closed system. Eating and breathing are both required to bring more energy into the system. The earth is also not a closed system—energy from the sun is constantly poured onto us.

The second law says that whenever energy is used to do work, some of it must be wasted and lost. You can't recover all the energy you use to move something, and use it again. There's a term for this wasted energy: *entropy*. According to the second law, whenever you do something, entropy must increase. More and more energy must be wasted.

The third law says there is an absolute zero—that there is a temperature where the jiggling completely stops.

Let's go back to the second law. One of its side effects is that it creates an "arrow of time." Entropy always increases. However, Newton's laws about how things move (and Einstein's, even though those wouldn't come for another hundred years) don't have an arrow of time. In those equations, if I reverse time, everything works the same. Things bounce off each other, and planets turn.

Temperature comes from atoms jiggling. Can't I undo entropy by having them jiggle the other way?

Also, the first law—that energy can neither be created nor destroyed—and the second law seem to contradict each other. If energy can't be destroyed, and I can't recover it, where does it go?

Are Newton's laws wrong, or is something else going on?

To tackle that we'll need to move from baking to mixing.

CHAPTER ELEVEN

ENTROPY EXPLAINED WITH MIXING

The first step in our recipe instructs us to mix together the flour, salt, and baking soda. Nothing has really changed about those ingredients— if you look closely in the bowl, you can still see the individual bits of each of them.

But at the same time, something fundamental *has* changed. You could theoretically unmix them—go through and separate the mixture back into a pile of flour, a pile of salt, and a pile of baking soda. But it would be difficult and take a lot of energy to do it. Mixing ingredients together is a one-way trip for all practical purposes.

One of the key things about Newton's and Einstein's laws of motion is that they are the same whether time runs forward or backward. If you took a movie of the planets moving around the sun and ran it backward, you would not be able to tell time had been reversed.

But this simple example of mixing our powder ingredients together shows a different truth. There is an "arrow of time." Explaining this arrow has led science to some deep and interesting places.

* * *

To help understand things, scientists like to develop simple models. For our model, let's use a bunch of cookies—frosted cookies in this case.

Each cookie can be in one of two states—either frosting down or frosting up. At the start, let's put all our cookies faceup.

This is a highly organized state and is comparable to when we first add the flour, sugar, and the other dry ingredients to the bowl. The ingredients are all clumped together, like all our cookies being faceup.

To simulate mixing, let's play this game: Pick a cookie at random and flip it over. So if it's frosting up, turn it to frosting down. And if it's frosting down, turn it back to frosting up.

Then we just repeat this step over and over again.

At the start, of course, when we pick a cookie at random, it will always be faceup since they all start faceup, so after the first step, we'll have one down and the rest up. On the next step, most likely another cookie will also be turned facedown, but it is possible that the first cookie we flipped will be turned faceup again. But gradually, as we do this over and over again, we'll end up with about half up and half down.

When we flip cookies, it is similar to mixing ingredients together. The order that we started out with gets lost.

Let's say I have ten cookies, and I go through this flipping process one hundred times. What does that look like? I went ahead and ran a simulation of this, and here are the results:

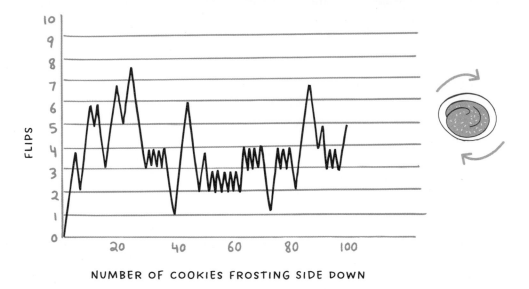

NUMBER OF COOKIES FROSTING SIDE DOWN

Each time you do this experiment, you'll get a different curve, but it will follow the same general outline. It starts at "one" (since you'll always have one face-down cookie after the first flip), and then rapidly rises until you get about half up and half down. Then it bounces around but generally stays close to half. In the above, if you look closely, you'll see that it does get as high as nine facedown cookies and as low as two. But it's mostly in the range of four to six.

Let's increase the number of cookies to one hundred, and flip them one thousand times. What does the curve look like now?

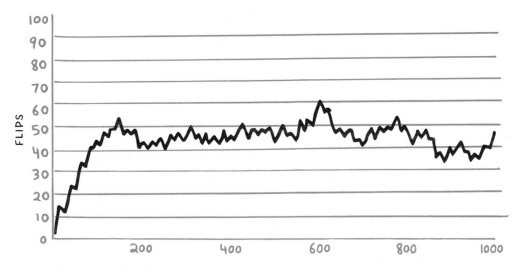

NUMBER OF COOKIES FROSTING SIDE DOWN

It looks similar—quick ramp-up, then it stabilizes. But once it reaches 50 percent up/down, it stays in a much narrower range. The lowest is thirty-eight, and the highest is sixty-four.

Let's go ten times bigger again: one thousand cookies. Here's the graph:

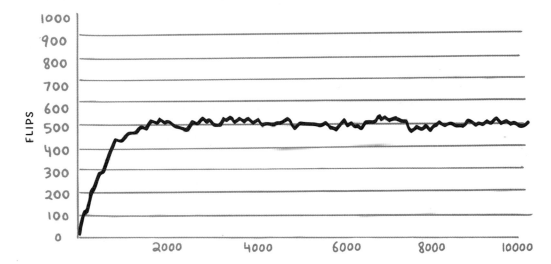

NUMBER OF COOKIES FROSTING SIDE DOWN

Same thing—just more of it. Quick ramp-up to 50 percent, and then it stays within an even narrower range. We stay between 470 and 530.

Is it possible the disorder can return to the lowest possible value, with all cookies frosting side up? Yes.

Is it likely? No.

In the ten-cookie case, there's a reasonable chance. It should happen once every 250 flips or so.

But in the hundred-cookie case, the chances of flipping over the "correct" fifty cookies to get back to all faceup is 1 in 10^{29}—so if you flip one cookie every second, it'll take about thirty trillion years. That's about two thousand times longer than the age of the universe, so . . . a really long time.

My calculator refused to figure out how long it would take with a thousand cookies. It's probably around 1 in 10^{200}—a really, really, big number.

When we combine flour, salt, and baking soda into a mixing bowl, the numbers are even more staggering. We got these huge numbers just by using one thousand or even one hundred cookie "particles." How many particles of salt, flour, and baking soda are there in the bowl? Way, way more than one thousand. So when you mix them, there are going to be many, many arrangements that have an average level of disorder—and only a few where, for example, all the salt is together.

So is it possible that after you thoroughly mix the ingredients you find all the salt in one spot? Yes. But it's incredibly unlikely. If you flip one hundred cookies, it takes 10^{29} years' worth of shuffles to maybe return to its starting position, so your bowl of ingredients will take many, many times longer.

* * *

This is the answer to the puzzle we posed earlier in this chapter. The laws of thermodynamics say that systems will gradually become more and more disordered. And yet Newton's laws say that any motion can be reversed. If all the billiard balls are moving the right way, they will re-form back into their starting rack. If you stir the flour, baking soda, and salt together in just the right way, they will separate back out. But, again, it's incredibly unlikely.

The second law of thermodynamics is not actually an ironclad law. It is based on probability. Entropy increases because it's a lot more likely that it will increase. It could decrease, but it is very unlikely, and not something that can be counted on.

If you look again at the graphs I put together on flipping cookies, notice again how much smaller the "ripple" gets when we move from ten to one hundred to one thousand cookies.

The objects we deal with in our world, and the planets and stars, are made up of astronomically many more particles. There are about ten thousand grains of salt in a teaspoon and about 10^{22} molecules of NaCl. So there is going to be basically no "ripple" in their random behavior. Randomness is still behind the scenes, but the huge number of particles averages it out.

Going back to Chapter Five about milk and cookies, that's why individual molecules can follow the weird rules of quantum mechanics, but a teaspoon of salt is perfectly well-behaved.

The second law of thermodynamics states that entropy always increases in a closed system, where no energy is added. However, if energy is put into a system, it can decrease entropy. We saw this last chapter with a pot of boiling water. Convection currents carry the water from the bottom of the pot to the top. It cools off there and drops back down to complete the cycle. This organized motion in the pot has lower entropy—more organization—than water that is just sitting there. The molecules can be organized in fewer ways to create those convection currents, thus the entropy is lower.

This doesn't violate the second law because you need to look at the whole system. Burning natural gas to create the flame that heats the pot increases entropy—and increases it by more than it decreases in the water. So the larger system—water plus flame—does increase.

Each hour, Earth is hit with about 173,000 terawatt-hours of energy. Humans as a whole consume about that in an entire year, so that's a lot of energy.

Much of that energy is radiated back into space. Some of it is absorbed by the atmosphere and creates weather patterns—the jet stream, storms, monsoon winds, and other convection currents. All of this is the earth using that energy to reduce entropy. Weather is a more organized system than the atmosphere just sitting still moving around randomly.

But the earth has the greatest entropy-reducing system we know of—life. Life is a constant struggle against disorder. The energy from the sun (and some from volcanic vents) allows living creatures to push back chaos, organize, grow, and reproduce.

Life, in a way, is like a hurricane—we take the energy from the sun to create and maintain our swirling pattern. But we are a much more intricate and beautiful pattern.

Life is a thin skin on our planet that turns sunshine into a kaleidoscope of form and function. Taking a step back and looking at the planet as an entire system, pulsing with the energy from the sun, is a magical perspective.

CHAPTER TWELVE

CHAOS EXPLAINED WITH VANILLA

Linking causes and effects to predict the future is a key part of science. But vanilla can illustrate how challenging that can be, and what that tells us about the universe.

The African nation of Madagascar is the top producer of vanilla in the world, purely due to the cleverness of a twelve-year-old enslaved person. Sometimes small things can have big consequences.

Vanilla beans are produced by an orchid native to Central and South America. It was introduced to other continents with the conquest and colonization of Aztec lands by Europeans. But unlike chocolate, also a product of the Americas, vanilla took much longer to become popular around the world. It wasn't until the 1700s that it started to be mixed with chocolate drinks in Europe, became a favorite of Queen Elizabeth I of England, and a delicacy in France. Thomas Jefferson experienced it for the first time during his time in France and handwrote a recipe for vanilla ice cream to bring back to Virginia with him.

As its popularity exploded, so did the price—because vanilla is very particular about producing its beans. The flowers open for only twenty-four hours before dying, and they can only be fertilized by a few specific pollinators.

Mexico was the main producer of vanilla in the 1800s, and many entrepreneurial botanists tried to transplant it to other regions. However, natural pollination didn't occur in those areas—the bees were wrong—and hand-pollination was laborious and time-consuming.

Enter Edmond Albius, a 12-year-old enslaved person. Edmond lived in Réunion, an island in the southern Indian Ocean just east of Madagascar. French colonists had brought vanilla plants there in the 1820s, but now, in 1841, they still were not producing vanilla beans in any meaningful way.

One day Edmond took what he had learned about hand-pollinating watermelon and carefully examined vanilla flowers. He realized that the pollen receptors in the vanilla flower were actually protected beneath a "lid" called the *rostellum*. And then he figured a way, with a small stick and simple flick of the thumb, to quickly and reliably lift the rostellum and pollinate the flower.

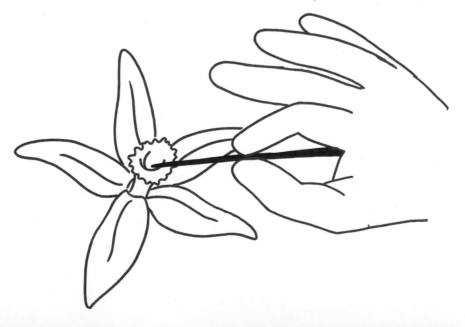

He showed Ferréol Bellier-Beaumont, the plantation owner, what he had discovered, and Bellier-Beaumont was amazed. He had Edmond demonstrate the technique to other enslaved people and other plantations around the island. Productivity on vanilla production skyrocketed, and less than a decade later, Réunion was the top producer of vanilla in the world. Bellier-Beaumont granted Edmond his freedom, but he never received any financial gain from his discovery, dying penniless. A statue was finally erected in his honor on Réunion in 1980.

No one could have predicted that Edmond Albius would define the economy of Réunion for hundreds of years. But part of the goal of science is to build models of systems that allow us to make predictions. If I know the current position and speed of a star, planet, and moon, I can figure out very accurately where they will be in the future. This is a very simple system, though, compared with trying to predict the course of human history (although pundits keep trying, of course). In between these two extremes are systems like fluid flow and galactic formation that computers can partially simulate. As computers become more powerful, our ability to probe these systems has become increasingly nuanced.

Studying the relationship between cause and consequence in increasingly complex systems has made us realize that our predictions can take us only so far.

It all started with the weather. When computers were developed, some of the first research performed with them was attempts to create models of the atmosphere and better predict the weather. Satellites fed much more data into these models, and were a big leap forward.

Models have continuously improved over time, and we now have fairly accurate weather forecasts. Meteorologists are pretty much spot-on for the next day or two, and usually fairly close up to five days out. Once you get from seven to ten days, though, things get a bit dicey. Why is that?

Computer models take data about the current state of the atmosphere and run through calculations to predict what will happen in the future. They process the model in small chunks of time—an hour on the most sophisticated models.

So if you input data for ten AM, the computer determines what the atmosphere will look like at eleven AM. The eleven AM data then gets fed into the model to determine what it looks like at noon, and so on.

Each step reads in information, processes it, and then spits it out to be read in again. And each of those steps introduces a little bit of error. As the steps proceed, those errors grow and grow as they build on one another, making the model less and less accurate the more steps there are and the further into the future it is simulating.

A side effect of this is that small changes to the initial data can lead to large changes in the output of the model. This sensitivity of computer models was first noticed by Edward Lorenz in 1961. He was running simulations of weather patterns, and the computer would print out all the data after each time step. He wanted to retest a run from a certain point in the middle, rather than run it from the start, since computers were slow and their time was expensive. So he reentered into the model the data from one of the time steps that had been printed out.

The results were close to the original model at first, but as the time steps went on, the prediction got further and further away. After a lot of digging, he realized the data inside the computer was tracked to six decimal places. But the printout had only three. So what was being stored inside the computer as 0.732319 printed out as 0.732. When he reentered 0.732 into the model, it ultimately gave completely different results.

Lorenz had discovered that a small change in the initial conditions could lead to a huge change later on. As he said about the results in his seminal paper:

"...if the theory were correct, one flap of a sea gull's wings would be enough to alter the course of the weather forever. The controversy has not yet been settled, but the most recent evidence seems to favor the sea gulls."

Others later convinced him to change this to a butterfly instead of a seagull, perhaps after the popular 1952 Ray Bradbury short story "A Sound of Thunder," in which a time-traveling dinosaur hunter accidentally steps on a butterfly, changing the history of the future. Whatever the reason, the change to a butterfly stuck, and this is now well known as *the butterfly effect*—small changes can have big impacts.

Lorenz's work helped create a new branch of mathematics—*chaos theory*.

The word *chaos* is used in a very specific way here. When we think of chaos, it is something completely out of control and can't be predicted.

But chaos in the mathematical sense means the rules that explain how a system develops over time are completely known and can be calculated. But despite that, if the start position varies just a tiny bit, the outcome changes a lot. And since we can't know the starting position 100 percent, we can't predict the outcome.

There is no randomness in this type of chaos. It is theoretically knowable, but practically is not. Spinning a roulette wheel is a type of deterministic chaos. If I precisely knew the force that was on the ball, the exact geometry of every surface on the wheel, and so on, I could work through Newton's equations and trace the precise path of the ball and where it will end up. But if any of those inputs is slightly off, it can change the path radically. If a dust particle lands on the wheel, it may slow the ball enough to hit a bump at a slightly different angle, and the error grows and grows, exactly like what was seen in weather models.

Chaos the way scientists think of it has two features:

* It is *deterministic*—if you repeat the same steps again with the same starting point, you end up in the same place.

* It is *unpredictable*—small changes in the start position lead to big changes in the end position, and often it can be difficult to tell what is going to happen in the future.

These seem incompatible, but it turns out that many, perhaps most, of the things we encounter in the real world act this way.

Lorenz himself built a simple device that gave a wonderful demonstration of chaos. It also allows us to bring back our old friend flour from the first chapter. Water-powered mills were a popular way to turn grain into flour for thousands of years, using the power of flowing water.

Lorenz constructed a waterwheel that had a series of buckets mounted around the edge, in a circle. But he put a hole in each bucket that slowly drained the water.

When the water pipe over the top bucket is turned on, the wheel will begin to turn. As buckets fill and empty, it will speed up, slow down, and reverse direction, all seemingly at random. Lorenz tracked the positions of the buckets, and there was absolutely no pattern that could be used to predict what it would do in the future. These waterwheels are wonderful to watch, and several have been installed as fountains and sculptures. An internet search on "Lorenz waterwheel" videos is a very pleasant way to spend some time.

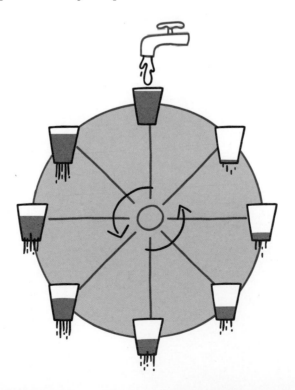

The Lorenz waterwheel and weather systems are pretty clearly challenging to calculate. The atmosphere is huge and complex, and even a wheel of buckets has lots of variables to consider—the angle the water hits the buckets at, the rate the water leaves the buckets, and so on. So in one sense, that these systems are chaotic is not that surprising.

But even simple systems can be chaotic.

For example, pick a number. Then:

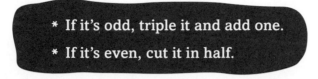

* If it's odd, triple it and add one.

* If it's even, cut it in half.

Then repeat those rules with the new number, and again, and again, and see what happens.

For example, let's start with the number seven.

It's odd, so we triple it and add one.

$$3 \times 7 = 21, +1 = 22$$

Twenty-two is even, so we divide by two

→ 11 .

Eleven goes to thirty-four, and so on:

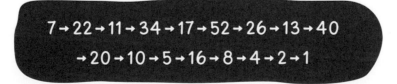

$$7 \to 22 \to 11 \to 34 \to 17 \to 52 \to 26 \to 13 \to 40$$
$$\to 20 \to 10 \to 5 \to 16 \to 8 \to 4 \to 2 \to 1$$

One goes back to four, and then we repeat the four, two, one loop forever.

Try this game with a few different numbers. You'll see that they all end up back at one. Usually, you'll bounce around between getting bigger and smaller.

This bouncing effect led mathematicians to call this a *hailstone sequence* after the way that hail can go up and down in a storm before it hits the ground.

Seven takes sixteen steps to reach one, and the biggest number it reaches is fifty-two.

But the number twenty-seven takes 111 steps (!) before it reaches one, and the highest number is a whopping 9,232 before it comes back down. Start at twenty-eight or twenty-nine, though, and it only takes eighteen steps before you get back to one. And twenty-nine just gets up to eighty-eight.

Two questions immediately come to mind:

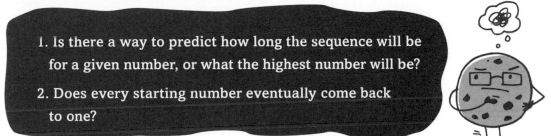

1. Is there a way to predict how long the sequence will be for a given number, or what the highest number will be?

2. Does every starting number eventually come back to one?

The answer to the first question is no. This hailstone sequence is an example of chaos. Small differences in the starting situation will change what happens dramatically, and unless we go through the steps, we don't know how it will turn out.

The answer to the second question is—we don't know. This is a famous unsolved problem in math called the *Collatz conjecture*, and mathematicians have spent a lot of time trying to prove or disprove it without any success. We just don't know.

You may have noticed a few paragraphs ago I suggested you try it with a few numbers and said it would return back to one. How did I know it would return to one if we haven't proved the Collatz conjecture yet? Strictly speaking, I didn't, but we have tested every single number up to 2^{68}, which is really, really big. In fact, it is:

295,147,905,179,352,825,856

If you happened to start with a number larger than that, and it never got to one, congratulations—you're famous!

Simple rules can absolutely lead to chaotic results, and chaos theory clearly shows us there are things that are unknowable and unpredictable. Chaotic systems just need to play out for us to learn their secrets. And layering on top of that the true randomness of quantum systems (discussed in Chapter Five), we can never precisely know what will happen.

But that adds its own charm to life. It even makes it possible for vanilla and a 12-year-old to change the course of history.

CHAPTER THIRTEEN

COMPLEXITY EXPLAINED WITH COOKIE CUTTERS

Perhaps the most common type of holiday cookie, particularly if you're baking with kids, is the humble sugar cookie. You roll out the dough and use cookie cutters to punch out fun shapes.

If you're anything like me—and I hope for your sake you're not—you like to play a little game when using cookie cutters. How many cookies can you get out of a sheet of dough? How can you arrange the cookie-cutter shapes in such a way as to have as little wasted dough as possible?

Yes—I know you can take your leftover dough, squish it together, and roll it again for more cookies. But let's ignore that for now. How easy is it to determine the best way to arrange your cookie cutters for maximum efficiency?

Well, mathematicians have studied this problem. And it turns out it's hard. Like really, really hard. Even with simple shapes.

For example, let's look at squares. Say I have a cookie cutter that makes one-inch square cookies, and I'm trying to get a specific number of cookies. What is the smallest square piece of dough that will allow me to get that number of cookies?

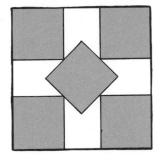

5 SQUARE PACKING

Well, if you want four cookies, the answer is simple. You just take a square that is two inches per side, and you can perfectly get four squares, with no wasted dough. If you go to five, the solution is to take your fifth square and rotate it 45° so it looks like a diamond. Then put the other four squares in the corner just touching the central diamond. The length of each side is about 2.7 inches.

So far so good—but the problem gets super hard super fast. For example, the best known arrangement for 11 squares looks wild. The squares are all at weird angles, and there are tiny gaps between some of them so that the corners of other squares can poke in between them. And it is not even known if this is the best arrangement!

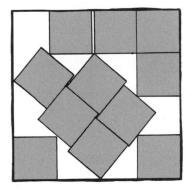

11 SQUARE PACKING

Because the squares can be moved and rotated minute amounts, these are almost impossible problems for a computer to solve. It is kind of wonderful that a simply stated and understandable problem like this is so difficult to figure out.

There are many famous problems, like the cookie-cutter problem, that people are working on. Here's another one.

Welcome to my bakery! I have for sale a lot of varieties of cookies. You are a true cookie connoisseur and have assigned a deliciousness value to each cookie. Each cookie also has a price, and since it is the end of the day and my bakery is a huge success, I have only one of each cookie left.

Here are the cookies I have and how delicious you rate them.

COOKIE	PRICE	DELICIOUSNESS
SHORTBREAD	$1.00	50
SNICKERDOODLE	$1.25	52
BLACK AND WHITE	$1.50	59
SUGAR	$1.75	60
MACAROON	$1.75	61
DOUBLE CHOCOLATE	$2.00	62
CRINKLE	$2.00	65
THUMBPRINT	$2.25	67
CHOCOLATE CHIP	$2.50	70
PEANUT BUTTER	$2.75	71
OATMEAL RAISIN	$3.00	76

You have three dollars to spend on cookies. What cookies should you buy that will give you the greatest total deliciousness? Take a minute, and see what you would do.

This problem is pretty simple. If you get one cookie, you want the oatmeal raisin, which gives you deliciousness of seventy-six. If you get two cookies, it is pretty easy to see that the best combo is the one-dollar shortbread and two-dollar crinkle, with a total deliciousness of 115 (fifty for shortbread, and sixty-five for crinkle).

What if you have four dollars? What is the best combination to buy then? Remember, I have only one of each cookie left. You could get the shortbread and oatmeal raisin, for a total of 126. But can you do better? You can, but I will leave that up to you to figure out.

Hopefully you see that just increasing the amount from three dollars to four dollars, which increased the number of combinations, makes this problem much harder. What if I increase it to twenty dollars? Or increase the number of types of cookies to twenty?

A characteristic of problems like this is that if someone hands you a potential solution to the problem, it is easy to tell if the solution is valid, but it is hard to tell if it is the best. For example, in this bakery problem, it's easy to tell if the cookies I pick cost twenty dollars or less. But is it the most delicious combination? That's a much harder problem. Similarly, with the cookie-cutter problem, it's easy to tell if the right number of one-inch squares are in the area but hard to tell if it's the best way to do it.

When we get above a handful of cookies, the number of possible combinations we need to consider explodes. There are hundreds of thousands of combinations of cookies you need to consider.

But some problems, like sorting a list of names into alphabetical order, do not explode as the number of names increases. It is harder to sort more names, of course, but the difficulty grows slowly.

The bakery problem is more commonly known as the *knapsack problem*, because usually it is presented as trying to pick what items you're going to put into a backpack that have the highest value, while still fitting. Problems like this are called *NP-Complete*, which basically means that it's easy to tell if a solution is valid, but really hard to find the best one. As another example, let's say you are moving across mountainous terrain and want to find the lowest spot. You might try to find it by moving downhill all the time. Eventually you'll reach a point where there's nowhere to go but up and decide that is the lowest. But it might not be, of course. It could be the lowest point in that region, but there might be a lower point somewhere else. The only way to be sure is to check everywhere.

The connection between complexity and chaos is clear. In a chaotic process, like the hailstone sequence we played with last chapter, there is no general shortcut to figure out how many steps there will be. You just need to crank through it.

Complex problems like these are the basis for internet security and encryption, using puzzles that take a long time to solve. When you're buying cookies online and enter your credit card information, it needs to be sent in a way so that someone intercepting the message can't see what the numbers are. They need to be put into a code and sent in an *encrypted* format.

★ ★ ★

Most of the puzzles used for encryption on the internet can be described as organizing piles of cookies. The puzzle takes this form:

Here's a pile of 7,493 cookies. Can you arrange them so that they form a rectangle, where both sides are two or more cookies long? Every time I ask you, I put in a different number of cookies, of course. 7,493 is just an example.

For example, if I have a pile of twenty-one cookies, I can arrange them like this:

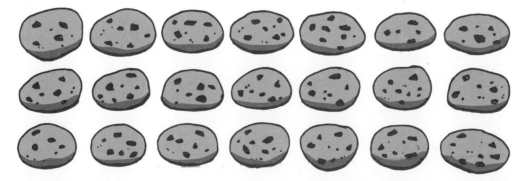

But if I give you a pile of twenty-three cookies, you can't arrange them in a rectangle. If you make a six-by-four rectangle, there will be one spot empty, and if you try a two-by-eleven rectangle, there will be one cookie left over.

If you can't put the cookies into a rectangle, you have a *prime* number of cookies. If you can, it's called a *composite* number of cookies. Sometimes you can make many rectangles—like if I give you twenty-four cookies, you can make a two-by-twelve, three-by-eight, or four-by-six rectangle. But twenty-five cookies can only be made into a five-by-five grid. The number of cookies on each side of the rectangle is called a *factor* of the number. So two, twelve, three, eight, four, and six are all factors of twenty-four. Five is the only factor of twenty-five.

If there's only one way to make the rectangle, it's going to be harder to figure out if the number of cookies you have is prime or not. Security and encryption on the internet are based on this idea. The root of the security is a really big number that one computer creates, that it knows is not prime, but has only two factors.

The computer on the receiving end of the message knows one of the sides of the rectangle (one of the factors), so it is easy for it to calculate the other factor, which is the secret value the first computer is trying to hide. But if a third person gets the message, it will take them a very long time to figure out how to arrange those cookies into a rectangle.

For example, it may not be obvious to you that 7,493 cookies can be arranged into a rectangle that is 127 x 59 cookies. But if you know that one side is fifty-nine cookies long, it will be very easy for you to figure out that the other side is 127.

This is a bit of an oversimplification of encryption and security, but it should give you the general idea of how it works.

The key point is that it is very important to the security of the internet that no one can figure out how to easily turn really big numbers of cookies into rectangles. If that complex problem becomes quickly solvable, that will cause big problems.

Unfortunately, researchers have figured out how to do that—theoretically at least—using something called a quantum computer.

* * *

The fundamental element of a computer is the bit—a value that is either a one or a zero. The fundamental element of a quantum computer is the *qubit*, short for *quantum bit*. Unlike a traditional bit, a qubit is not just a single value—a one or a zero. Since we are in the quantum world, it has a probability of being a one or a zero. As you may recall from Chapter Five, particles under the laws of quantum mechanics do not have a single definite position. They are represented by *probabilities* that they may be found in certain positions. Similarly, a qubit is a probability, but when you measure it, you always measure it as a zero or a one.

Because qubits represent multiple values, researchers have figured out that if they can connect multiple qubits to one another in specific ways—called *entangling*—then the qubits can find the factors for very large numbers much, much faster than any traditional computer.

However, the technical hurdles for getting qubits to work reliably are daunting. Because quantum effects are so tiny and preserving the delicate relationships between the particles is so critical, qubits need to be kept close to absolute zero, and they need to be kept away from other particles. Even so, there is always a chance for error, so additional qubits need to be used to detect and correct errors, which makes the whole system that much more complex to design.

So far, qubits have been used only to determine very simple factors—showing, for example, that twenty-one is three-by-seven. This is obviously not that impressive on its own, but it was an important milestone for showing that the theory works. The best experts in the field predict it will take about ten or twenty years before quantum computers start to reach their full potential.

Even though qubits sound amazing, it is important to understand that quantum computers are great at solving only certain types of problems—like putting cookies into rectangles or figuring out which cookies to buy at our single-cookie bakery. But they are not great at everyday computing problems. The computers we have today are terrific at that. Quantum computers will be specialized devices for special purposes.

Complexity and chaos are two sides of the same cookie. You may notice that they both limit how quickly and precisely we can understand something or predict the future. There's one more way to look at the complex cookie of chaos (the chaotic cookie of complexity?), by taking these concepts and applying them to shapes that look simple but get more complicated the closer you look. Let's use an oatmeal raisin cookie to explore the world of fractals.

CHAPTER FOURTEEN

FRACTALS EXPLAINED WITH AN OATMEAL RAISIN COOKIE

While chocolate chip is my number one cookie, oatmeal raisin is right up there. The texture, the chewiness, the flavor—it's a great package. And they can help us pull together the strands of chaos, complexity, and uncertainty that we've explored the last few chapters.

Here's a picture of an oatmeal raisin cookie:

You may remember from your early schooling how to determine the perimeter of an object. If you don't remember that word, it's how long the outside edge is. For example, the perimeter of this rectangle is sixteen (5 + 3 + 5+ 3):

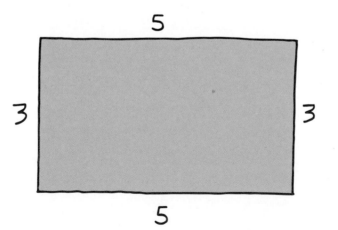

How would you figure out the perimeter of an oatmeal raisin cookie?

For your first guess, you might draw a series of lines that basically outline the cookie and measure them. Something like this:

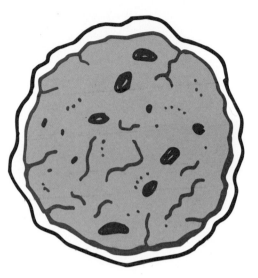

If you shrunk yourself down really small and wanted to walk around the cookie, you might follow those blue lines.

But those lines are obviously not exact. Let's zoom in on the upper left of the cookie. See how it's more crinkly than it looked from far away?

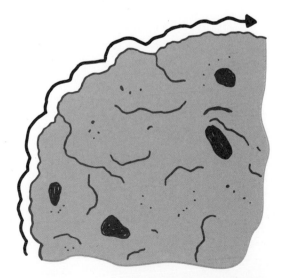

We need to add more lines to get closer to the actual border.

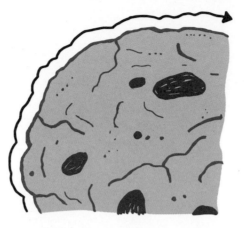

If we measure the blue lines we drew, the border is now longer. If our shrunk-down-self walked around this new border, it would take us longer to get all the way around. We're spending more time going in and out of nooks and crannies.

But if we zoom in, again it looks even more crinkly. And we have to add more lines again, making our perimeter and our cookie-walk even longer.

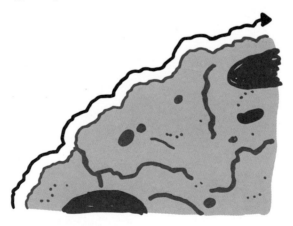

In fact, as we keep zooming in more and more, we see more and more crinkles around the edge. Smaller and smaller crinkles, of course. But we never reach a point where we'll just see nice straight lines that we can measure.

The edge of our oatmeal raisin cookie is theoretically **infinitely long**. The *area* of the cookie is finite—we can draw a circle that's slightly larger than the cookie, so we know the area has to be smaller than that—but the perimeter just wiggles back and forth, getting longer and longer.

This type of shape is called a *fractal*.

One of the simplest shapes that, like an oatmeal raisin cookie, has a finite area but an infinite border is called the *Koch snowflake*, discovered in 1904 by Swedish mathematician Helge von Koch.

To build it, start with a triangle. Take the center third of each edge, bump it out into another triangle, and then repeat. Here are the first three steps:

If you keep repeating this over and over again, the perimeter gets bigger and bigger—and never stops increasing in size. It continues on to infinity. The area, though, will converge to 1.6 times the area of the first triangle.

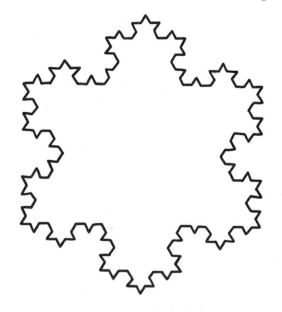

The idea of repeating the same steps over and over again should be familiar. It was how we generated chaos with our weather models. Computers are really good at it, which is why research into fractals exploded as computers became more powerful.

While for many points it's obvious if they are inside or outside the shape, for points that are right on that crinkly border, it is impossible to tell if the point is in or out. You have to actually run through the steps to figure it out. You can't tell exactly where you are until you grind through the calculations.

The snowflake (and our cookie) are *self-similar*. As you zoom in, it keeps looking the same. This is a common characteristic of fractals and is one of the hallmarks of perhaps the most famous of fractals, the *Mandelbrot set*, named after Polish mathematician Benoit Mandelbrot, who pioneered and popularized work on fractal geometry.

I won't go into the details of how the Mandelbrot set is constructed, as it gets a little mathy, but it's pretty straightforward and there are many tutorials on the internet if you're interested. But the resulting shape is mesmerizing. As you zoom in on the edge, you find swirls and spirals of increasing delicacy, and even miniature copies of the Mandelbrot set shape itself. It is a visual feast.

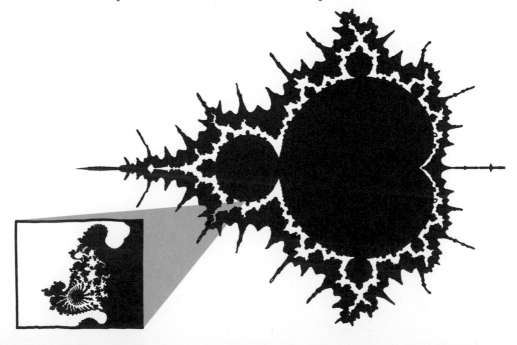

While fascinating to look at, fractals are very representative of nature—more representative than classic circles and squares. Trees, lightning, clouds, mountains—and cookies—are all fractals. Signals can be fractals as well—our heartbeat is fractal in nature. An EKG gets more "crinkly" the more you zoom in on it. In fact, a heartbeat becoming less fractal—smoothing out—is an indicator of an increased risk of heart disease.

The last few chapters—chaos, complexity, and now fractals—all point at a deeper underlying truth about nature and our limitations.

Science goes back to a core function of our brains—trying to make predictions to keep us out of trouble by making observations about what is happening in the world.

We observe, we predict, we make connections between cause and effect.

When Newton developed his laws in the seventeenth century, our ability to predict how things would move took a great leap forward. According to Newton, if you knew where everything was and how fast it was moving, you could predict what would happen—forever. This became known as "the clockwork universe."

This idea of the clockwork universe, of everything moving in a great cosmic ballet, became even more enshrined through the eighteenth and nineteenth centuries, as more discoveries were made that fit this idea. Electricity and magnetism were measured and tamed and found to follow a set of laws similar to Newton's laws of motion, called Maxwell's equations. Maxwell's four equations told you exactly what would happen with electromagnetic fields for all time; they also showed that light was just a wave of electric and magnetic energy.

Atomic theory showed how the elements could be organized and understood, and genetics and evolution gave an organizing principle to biology.

By the end of the nineteenth century, it seemed realistic that the program of science was drawing to a close—that we soon would have at our fingertips rules that explained how everything works, and that would allow us to predict, with stately elegance, what would happen.

But through the twentieth century, that vision shattered.

Chaos theory showed that even in systems with well-defined rules, you can't always predict the future.

Complexity theory showed that certain problems simply could not be solved without just going through the options, and those options exploded so quickly that some problems basically become unsolvable.

The discovery of fractals demonstrated that geometry could get weird and fuzzy and also impossible to predict.

Quantum mechanics showed that we could know only the *probability* of the different things that might happen. We could not predict a single future with certainty. The locations and interactions of particles are impossible to know for sure.

* * *

Even math and logic are not immune to this encroaching chaos and complexity.

The idea of the math "proof" goes back to Euclid over two thousand years ago. He developed the method of proof we still use today—start with *axioms*, statements we agree are true, rules on how we can combine axioms, and figure out a way to go, step by step, from an axiom to the statement we're interested in. This allows us to figure out if statements are true or false—if they can be proved or disproved.

Euclid's axioms include statements like "you can draw a straight line between any two points" and "there is only one circle with a given center and radius."

This idea of proof is very powerful and has driven math discovery since Euclid. Mathematicians believed that with the right set of axioms, they could prove or disprove any math statement. That every math proposition was provably either true or false.

In 1931, that dream was shattered. Kurt Gödel proved (ironically) that in any mathematical or logical system, there would always be statements that could not be proved to be true or false. He called them *undecidable*. And you can't even know for sure if a statement is undecidable, or just really, really hard to prove or disprove.

The *Collatz conjecture*, from Chapter Twelve, is a possible example. We don't know if it's true or false. Mathematicians have tried to prove it or disprove it without success. But is it just because we haven't been clever enough yet to figure out a proof? Or is it because the statement is one of Gödel's undecidable propositions? We can't tell.

We do know that problems we thought might be undecidable ended up being solved. One of my favorites is the four-color map problem. It's easy to state—picture a page in a coloring book. What is the fewest colors you need to color the spaces so that no two spaces that touch are the same color? Touching just at a corner doesn't count. It has to be a side.

It is easy to show that three colors are not enough. Here's a design you can't color with just three crayons:

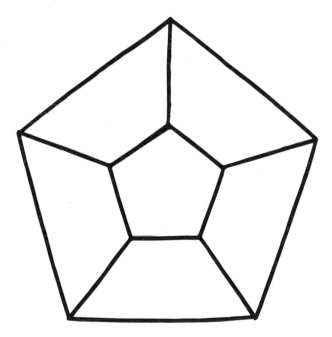

You need four colors for this. But is four colors always enough? Or is there some complex pattern that requires five?

This question was first proposed, as far as we know, in the 1850s. And various attempts were made to prove it over the ensuing decades, all failing.

After Gödel, mathematicians started to wonder—was this theorem undecidable? Would it never be proved or disproved? In this case, though, the theorem fell to computers. In 1976, a computer was used to prove that, yes, all you need are four colors in your crayon box, and you can color any map so that no two adjacent spaces have the same color. But it took over one hundred years to show that.

Is the Collatz conjecture provable or undecidable? If it's undecidable, we'll never know. We'll always think it *might* be provable but never be certain.

Logic itself has shown that the truth is a fractal. There will be statements we know are true, and those we know are false. But there will also always be statements that lie on a fuzzy boundary between the two.

With the Koch snowflake or an oatmeal raisin cookie, if I give you coordinates of a point, it may be definitely inside the boundary, or it may be definitely outside. But it may also be right on the edge, and no matter how far we zoom in, we can't quite tell precisely.

The universe has a fundamental level of chaos and complexity. Not everything is knowable. We will never be done learning new things. There will always be new discoveries to make and problems to unravel.

Personally, I find this as comforting as a plate of warm cookies.

CHAPTER FIFTEEN

EXOPLANETS EXPLAINED WITH A NICE GOLDEN-BROWN COLOR

Our cookie recipe instructs us to bake them until they are "golden brown." Much of our enjoyment of the world comes from colors. The brilliance of flowers, the beautiful plumage of tropical birds, the deep red of a setting sun, even the golden brown of a chocolate chip cookie— all speak to us as humans. But color will also help look beyond humanity as we search for intelligent life on other planets.

What is a color? In one sense, it is very simple—a color is light at a specific energy. Human perception of color is very complex, though.

We don't see the "color" of light directly. Receptors in our eyes respond to red, blue, and green light, and our brain combines these to create colors, like the golden brown of our cookie. For our purposes, though, let's focus on the colors that are carried by light.

Light is a little packet of electromagnetic fields bouncing up and down. While the light packet itself always moves forward at a constant speed—the speed of light, of course—it can vibrate at any frequency. And this frequency defines the color. There are two choices on how to describe a light wave. One is to measure the *frequency*, how many times per second the wave wiggles up and down. Another is the *wavelength*, which is the distance between the peaks of the waves. Since the speed of light is constant, either the wavelength or frequency is enough to tell you about the color of the light. For visible light, we most often use wavelength, measured in "nanometers" (nm). A nanometer is a billionth of a meter, so quite small. Blue light, for example, is about 500 nm, and red light is about 700 nm.

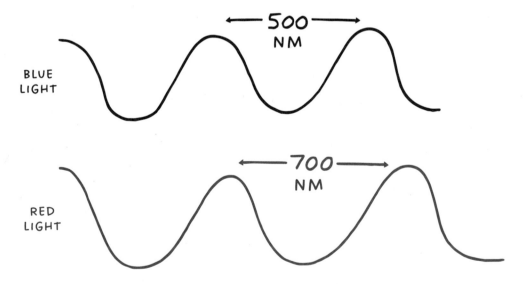

Newton was one of the earliest scientists to study light. He realized that white light is a combination of all the colors and that a prism could break it apart into a *spectrum*.

The visible spectrum contains an infinite number of colors, but Newton broke it into seven—red, orange, yellow, green, blue, indigo, and violet, mainly because he thought the number seven had special significance. But there's no reason at all that it needs to be seven. In fact, Newton at first decided it was six colors, but he threw in indigo at the last minute because he wanted seven.

The spectrum doesn't end with the visible portion. If you move past red, you get to infrared, microwaves, and then radio waves, as the wavelengths get longer and longer. If you go past the violet end, the wavelengths get shorter and shorter, and you get ultraviolet light, x-rays, and gamma rays.

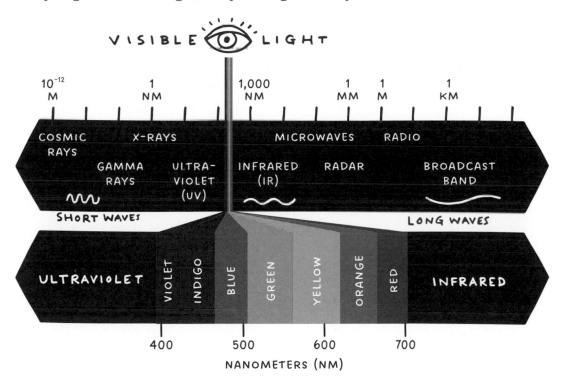

After Newton's time, prisms, telescopes, and optical equipment got better and better. In the early 1800s, scientists noticed that when they looked at sunlight through a prism, it wasn't actually a continuous spectrum. There were thin black lines seemingly at random throughout the spectrum. Some colors were just missing. Why?

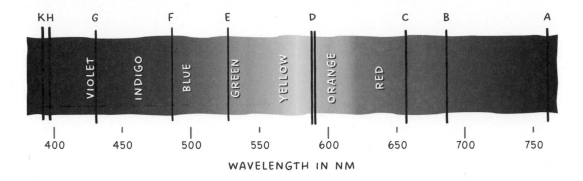

Fifty years later, some clues started to come into focus. Scientists noticed that different elements made flames of different colors when they were burning. Our old pal sodium, from salt, burns orange. Potassium burns pink, and copper green. Some elements, like neon, glowed when electricity was passed through them.

When they looked at the light spectra of these flames, they each showed a different pattern of lines. These lines form a fingerprint of the atom.

Scientists also discovered that elements only absorbed light that was the same frequency as what they emitted. If light passes through a cloud of mercury, lines appear at the same wavelengths.

When they went back and looked at the light coming from the sun, the black lines matched up with lines from certain elements. *The lines were telling us what the sun was made of.* As light emerges from the sun, it passes through those elements, and certain colors get absorbed.

Scientists were able to match up the lines from the sun with the lines we could re-create here on earth and figure out the elements.

Ever since we discovered that the sun is a star, we have wondered if the stars also have planets orbiting them, like our solar system. Planets outside our solar system are called *exoplanets*. We were pretty sure there must be exoplanets, but we weren't sure what percentage of stars had planets or what those planets looked like. How many of them were Earthlike? Was an Earth twin out there waiting to be discovered?

Stars are incredibly far away, so as you can imagine, it is very difficult to directly see a planet—particularly since they don't shine. Any light coming from them is reflected from their sun.

Astronomers use two primary ways to detect planets. One is to look at the light coming from the star. If the orbit of the planet is aligned to Earth, when the planet passes in front of the star, some of the light is blocked. If the dimming happens regularly, you can infer that there is a planet. The amount of light that is blocked, and the time it takes the planet to go around the sun can help tell you how large it is and how far away from its star.

In 2009, the Kepler Space Telescope was launched. Its purpose was to try to find exoplanets by continually monitoring stars, looking for dimming. Over the next nine years, the Kepler mission found thousands of stars that have planets around them.

For the second method, we will need to go back to something we discussed in Chapter One. As you may recall, gravity works both ways. Earth pulls on you, but you also pull on the earth.

Just as we pull on the earth, the earth pulls on the sun. Our usual picture of a solar system is that the sun is stationary in the center, and the planets revolve around it.

But that's not actually the case. If you look at an entire solar system from far away, you will see that the sun is actually orbiting an imaginary point called the *barycenter.*

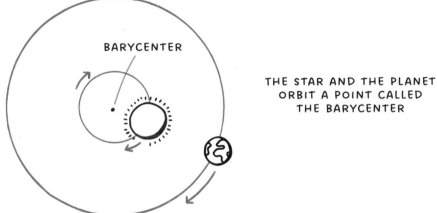

BARYCENTER

THE STAR AND THE PLANET
ORBIT A POINT CALLED
THE BARYCENTER

As the star wobbles, it moves toward us and away from us. If a star wobbles, we know that other objects are orbiting it. The size and speed of the wobble tell us something about them.

STELLAR WOBBLE

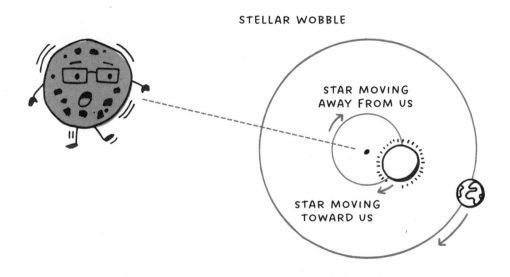

STAR MOVING
AWAY FROM US

STAR MOVING
TOWARD US

We can't directly see a star moving toward and away from us. It does not move that much. However, something else changes that is easier for us to measure—the color.

If you've ever stood near a train as it passes by you, or attended a car race, you know there's a characteristic sound as something quickly moves past. As it approaches, it gets louder, of course, but the pitch also rises. Then, when it passes by, the tone changes, and it gets softer and the pitch drops. This is particularly noticeable if a train blows its whistle as it goes by.

This is called the Doppler effect. Sound, like light, is a wave. If the train is moving toward you, the distance between the peaks of the wave shrinks, so the pitch gets higher.

Similarly, if the train is moving away from you, the wavelength will get longer, and the pitch will go down.

The same thing happens with light. If a light source is moving away from you, the *frequency* will shift downward. As you may recall from the spectrum earlier in this chapter, if visible light has a lower frequency, it is toward the red end of the spectrum. If the light source is moving toward you, the light frequency will increase, and it will shift toward the blue-violet end of the spectrum.

So we can look at the light from a star and see whether it shifts blue or red to determine how fast it is moving toward or away from Earth. But we have a problem. We need to know what the unshifted frequency is so we can see how much it shifted. How can we figure that out?

Well, you already met the answer. We can tell what the sun is made of by seeing the lines in the spectrum and match those up with the elements that produce them.

We know, for example, where the lines for hydrogen should be. When scientists see those lines in stars, if they are exactly where we expect, then the star is not moving relative to us. If the lines are farther in the red than we expect, it is moving away from us. And if the lines are shifted toward the blue, the star is moving toward us.

This is called redshift and blueshift. And the more something is redshifted, the faster it is moving away.

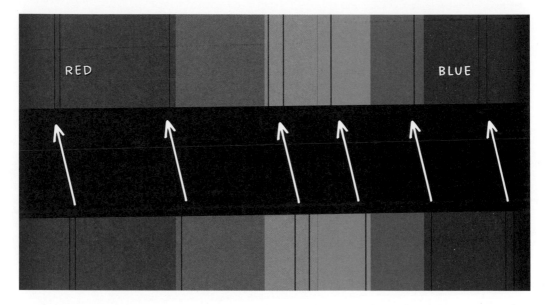

THE HYDROGEN LINES IN THE TOP SPECTRUM ARE SHIFTED INTO THE RED, SO WE KNOW THAT STAR IS MOVING AWAY FROM US.

A wobbling star cycles between moving toward and away from us. If we watch the spectrum coming from that star, we will see the hydrogen lines shift toward the blue as it moves toward Earth and then toward the red as it moves away.

* * *

Using these and other methods, we have now found over five thousand planets outside our solar system. We've only started to look and are still refining our techniques for finding them. The James Webb Space Telescope, which just became operational in 2022, has even allowed us to directly image larger planets directly.

Current estimates are that our galaxy has hundreds of billions of planets.

So we've found lots of planets outside our solar system. But how can we tell what the planets are made of? Are they rocky planets like Mars or gas giants like Jupiter? Or maybe they have water like Earth?

To determine if there might be water, we can look at how hot the star is and how far away the planets are from it. If they are very close to their sun, we can assume that the planets are so hot that any water they might have had has boiled away. If they are very far away, then any water would be locked away as ice. There is a "goldilocks zone" where liquid water may exist, which is called the *habitable zone*. Current estimates are that five billion to ten billion planets in our galaxy are in the habitable zone.

For a planet to be like Earth, it also needs to have an atmosphere, and hopefully composed similarly to ours. Could we tell what the atmosphere is composed of?

As you may have guessed, spectral analysis can help us out here. Many planets are detected by looking at how the light from the star dips slightly as the planet moves across its face. As it does so, a tiny bit of the light from that star passes through the planet's atmosphere on its way to Earth.

By comparing the normal spectrum from the star, with the spectrum of light that has passed through the planet's atmosphere, we can deduce what elements must be there. We can look at which new dark absorption lines appear or which ones are darker than the star alone.

As you can imagine, this is a very tricky measurement. The size of the planet is already small compared with the star, particularly for an Earth-size world. And the atmosphere is just a tiny slice of that. The precision required to do this analysis is impressive.

The first spectrum of an alien world was measured in 2001, when sodium was discovered in the atmosphere of a distant planet. Since then, a variety of atmospheres have been found, ranging from very exotic to very familiar.

So when you peek into your oven to check on whether your cookies have achieved that golden-brown color, think about scientists trying to catch a glimpse of the colors on planets light-years away.

CHAPTER SIXTEEN

THE BIG BANG EXPLAINED WITH CHOCOLATE CHIPS

We are now, at long last, ready for chocolate chips. They're the only ingredient in the name of the cookie, so they must be important. And, indeed, they are. Chocolate chips are our stepping stone to learn about the dawn and ultimate fate of the universe itself.

Imagine, if you will, that you have placed a flat disc of cookie dough onto the baking sheet. You shrink yourself down until you are way smaller than a chip and pick one to stand on.

As the cookie bakes, it expands in all directions and goes from a thick disc to a thin cookie spread out across the baking sheet. If you stand on a chip and look at the other chips, what do you see? It looks like all the other chips are getting farther away from you. And the farther away the chips are, the faster they are moving away.

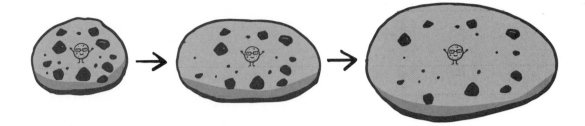

In the 1920s, astronomers started to notice a similar pattern in our universe. Galaxies all seemed to be moving away from us. And the farther away they were from us, the faster they were moving.

So what does that mean? If everything is moving away from us, does that mean we are actually at the center of the universe?

Before we get into that, there are two questions you may be asking yourself. How do you measure the distance to a galaxy? They are really, really, really far away. And how do you tell how fast it's going?

The simplest way to measure the distance to an object is *parallax*. You measure the angle to an object. Then you move sideways a known distance and measure the angle again. A little math will tell you how far away the object is.

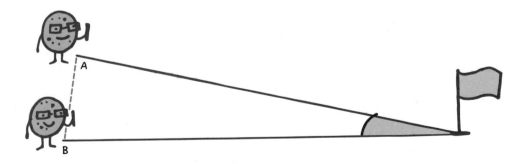

The farther away the object is, the more you need to move to get an accurate measurement. For astronomers here on Earth (which are all the astronomers we have right now), the farthest you can move is from one side of the sun to the other. If you take one measurement to a star in January and then another in July, you will have a baseline of about 180 million miles. With a base that wide, you can make a fairly accurate distance measurement to about 1,000 light-years away. Anything farther away than that and the angle change is too small to measure properly.

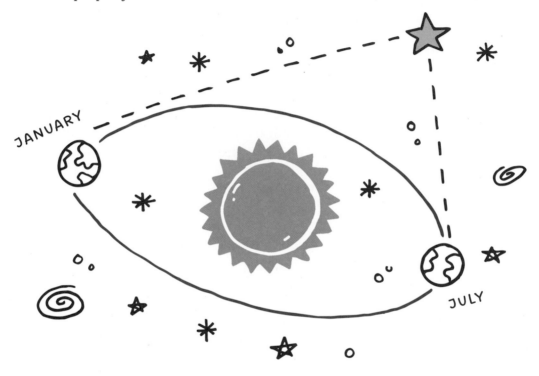

A *light-year* is the distance light travels in a year. Light travels fast, so a light-year is pretty far—about 5.8 trillion miles. But one light-year is a drop in the bucket, galactically speaking.

Our galaxy is 100,000 light-years across, and other galaxies are even farther away. So the parallax method is not going to cut it beyond our local stellar neighborhood.

How then do we measure longer distances?

One possibility is to see how bright something is. Let's say you look at your phone screen from one foot away. If you look at it from across the room, it will not appear as bright. As you move farther away, the screen gets dimmer and dimmer. By measuring the brightness, you can calculate the distance.

However, for this to work, you need to know how bright the star is. And stars vary greatly in their brightness. You can tell a little bit by the color, but it's hard to be accurate enough to be confident in the distance measurement.

What we need is a standard star—one where we know what the brightness is.

In the early 1900s, Henrietta Leavitt was a "computer" at Harvard Observatory. Computers back then were people who specialized in doing math calculations.

In 1912, Leavitt was analyzing variable stars in the Small Magellanic Cloud, a dwarf galaxy. Variable stars vary their brightness in a regular period, ranging from a few days to a few months. And she discovered something remarkable. The period was directly related to the brightness. The faster the star pulsed, the dimmer it was. The slowest stars were the brightest. And most important, this relationship was a straight line. If you knew the period, you knew the brightness.

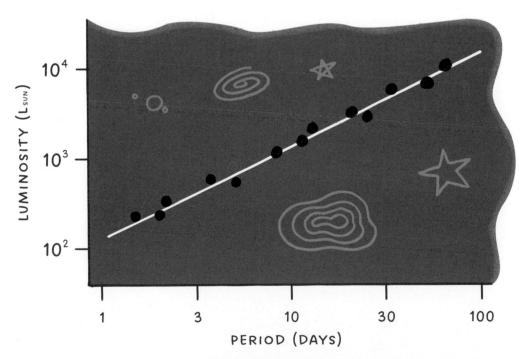

CEPHEID VARIABLE STARS

These special types of stars, named Cepheid variable stars, were exactly what astronomers were looking for. It gave them a "standard candle" that could be used to measure distance. When looking at a galaxy, astronomers look for Cepheids. They are common enough that they can almost always be found. And once you find the Cepheid and measure how many days it takes to blink, you know how bright it must be.

As astronomers measured the distances to more and more galaxies, they learned of many other ways to measure distances as well. By using a variety of methods, we think we have a pretty good idea of how far away most objects are.

If the galaxies in the universe are moving apart as space stretches, we can run the movie backward and look at the past. If you do, it shows that the universe would have been concentrated in a very small area about fourteen billion years ago. The Milky Way itself is probably around thirteen billion years old, and Earth about 4.5 billion years. Life has been on earth for about three billion of those.

The initial point in time, when the universe was crammed into a small space, is called the Big Bang. At first, several alternative theories competed with the Big Bang, but observations in the 1950s and 1960s made it clear that it was the best explanation for the data.

There were two particularly strong pieces of evidence.

First, as you look at galaxies that are farther and farther away, you are also—in a very real sense—looking back in time.

The light from the sun takes eight minutes to reach us. We actually see the sun as it was a few minutes before. The closest star—Alpha Centauri—is four light-years away. When we look at it, we see it as it was four years ago. The Andromeda galaxy is 2.5 million light-years away, so the light we see from there now started its trip as the first human ancestors walked the earth.

The farthest galaxy we've seen is a shade over thirteen billion light-years away. That means we're seeing light that started its journey before our sun even existed.

The Big Bang predicts that over the course of the fourteen billion years of the universe, stars and galaxies were born and evolved. And sure enough, that's what we see. In the most distant galaxies, we see *quasars*, incredibly energetic stars that formed in the early universe. Theory predicts that quasars don't last, and sure enough we don't see them in the galaxies that are closer to us—in other words, in the more recent history of the universe. They all died out by then.

For the other piece of evidence, imagine an oven. Let's say you warm up the oven to 500°F and then turn it off and track the temperature. It will gradually cool down. By understanding the oven, you can make a very good prediction about how the temperature will change over time.

At the start of the Big Bang, all energy in the universe was concentrated in a very tiny area, so the temperature was incredibly high. Over time, the general temperature of the universe has dropped.

In an oven, the temperature drops because the energy inside the oven gradually equalizes with the energy outside the oven. However, with the universe, there is no "outside." The temperature drops because of the stretching of space. If the same amount of energy is spread over a large space, the average energy will be lower.

So does this background hum, this leftover energy from the Big Bang exist? Yup! It was actually discovered by accident. Based on the estimated age of the universe, the energy should be in the microwave range, detectable by radio telescopes. When the first radio telescopes were developed in the 1950s, they constantly detected a consistent low-level signal. At first, researchers thought it was a defect in their equipment, that a loose cable or something must be interfering with the signal. But after pulling their hair out for several months, they finally realized that it was a real signal and that it matched the expected frequency of the cooled-down Big Bang radiation.

One of the amazing things about this radiation—the cosmic microwave background radiation, or CMBR for short—is how incredibly consistent it is. Wherever astronomers looked in the sky, it was there, and it was very even across the entire universe. The CMBR does vary slightly—but only to about one part in ten thousand.

However, even this slight variation is important. It is critical to our universe. The variation in the CMBR shows that energy wasn't exactly evenly spread out at the earliest points of the universe. And those variations—wrinkles, if you will—led to slight variations in the density of the first atoms and meant that there were areas slightly higher in gravitational force than other areas. This led to bigger clumps, which had more gravity, and so on, ultimately leading to stars and galaxies. And us.

That variation is behind everything we can see.

In the 1990s, astronomers launched the Cosmic Background Explorer (COBE), which mapped out the CMBR in great detail. The famous image it created is a fingerprint of the early universe.

Since the modern Big Bang theory was proposed in the 1950s, astronomers wondered about the final fate of the universe. Would the galaxies continue to move farther and farther apart forever? Or would the expansion reverse and everything start to move back together, the so-called Big Crunch?

One thing astronomers did know was that the expansion rate of the universe was slowing down. It had to be.

Back in Chapter One we talked about Newton's laws. His core idea was that an object would keep moving at the same speed unless acted on by a force. Well, galaxies are pretty simple from this standpoint. The only force acting on them in any significant way is gravity. It's the only force that is strong enough on the scale of the universe to have any effect.

All the galaxies pull on one another. And that means that the expansion rate should be slowing down, as the force on all the galaxies is in the exact opposite direction of intergalactic expansion. We weren't sure if it was slowing down enough that the expansion would reverse, but everything we knew about the universe said that the expansion had to be slowing.

But measurements taken over the last twenty years have shown that the expansion of the universe is not slowing down. It seems to be speeding up. Which is very, very weird. There must be some force or energy that is either pushing galaxies apart or causing space to stretch faster and faster.

Scientists have called this *dark energy*. We don't know what it is or where it comes from. Based on what we observe, however, we believe that it makes up more than two-thirds of the universe.

As you may remember, back in Chapter One we talked about *dark matter*, which is matter that exerts a gravitational force. We have no idea what dark matter is. Most of the promising theories have been shown to be incorrect. And dark matter accounts for about ¼ of the universe.

So, to summarize—

DARK
ENERGY

68%

DARK
MATTER

27%

5%

EVERYTHING
ELSE

(ALL THE PROTONS,
ELECTRONS, NEUTRONS,
PHOTONS, AND EVERYTHING
THAT MAKES UP PLANETS,
STARS, AND GALAXIES)

You may have noticed that this chart shows that Dark Matter is 27 percent of the universe. In Chapter One we said that it was 85 percent of the mass. That's because Dark Energy isn't considered part of the "mass" of the universe.

We don't know what dark energy is, and we don't know what dark matter is. We are currently completely clueless about 95 percent of what the universe is made of.

There are a few theories about what dark energy is. One is that it is simply a property of space. Quantum mechanics tells us that the vacuum of space isn't really a vacuum. Particles are constantly created and destroyed, in a roiling mass of activity. It is possible that this creates a pressure that forces space to stretch more than expected. However, calculations to try to project what this force would be are completely and totally off from what we observe. Like one-with-a-hundred-zeros times too big.

Perhaps it is a new type of substance, an energy field we don't understand yet. Scientists who champion this idea call it *quintessence*, a word used in the Middle Ages to describe a mythical fifth element, joining earth, air, fire, and water. However, we don't have much more than a name at this point.

Finally, it is possible that Einstein's general relativity gravity equations are incorrect. If so, it is possible that there really is no such thing as dark energy or dark matter. Both make what we see match our theory of gravity. But if the theory changes, it is possible these ideas go away as well.

For decades back in the 1800s, scientists believed that electromagnetic waves—light—had to travel through space in something they named the ether. All waves that we knew about up until then had to travel in something. Ocean waves travel in water, and sound waves need air. Scientists assumed that in a similar fashion, ether carried light waves. Ultimately, theory and experiments improved, and it was shown that light could travel just fine in a vacuum. The electric and magnetic waves that make up light support each other and don't require "shaking" something outside of the light itself. Similarly, changing gravity might make dark energy and dark matter unnecessary.

We do know for sure that our two most successful theories ever in physics—gravity and quantum mechanics—are not exactly right, because they don't agree with each other. Our theory of gravity works incredibly well at large distances. Every time we have tried to directly test it, it has proved to be correct.

And quantum mechanics does an amazing job at giving answers about what happens to the tiniest of particles. The predictions match experiments out to many, many decimal places. Yet our theory of gravity does not work at the tiniest scales and conflicts with quantum mechanics.

So there must be something missing. Our chocolate chip cookie is spreading out on the baking sheet faster and faster, and we don't know why. I am excited to learn the answer.

THE UNIVERSE EXPLAINED WITH A COOKIE

I left the chocolate chip for last for an important reason. What it represents—the life cycle of the universe—pulls together all the topics we have talked about. The ingredients and steps to bake a chocolate chip cookie set humanity on the path to comprehend the full scope of the universe.

Flour and sugar led to understanding gravity and galactic structure, salt and baking soda to the tiniest particles and the quantum world. Vanilla, cookie cutters, and walking around an oatmeal raisin cookie brought us to fundamental truths about chaos, complexity, and how much we can ultimately predict. Mixing and baking led to understanding thermodynamics and entropy, which just might underlie everything.

And brown sugar and measuring spoons showed us how to be more accurate in our measurements and appreciate and deal with errors that always arise. Color taught us how to build a measuring stick for distances that boggle the imagination. Finally, eggs, butter, and a bit of cookie decorating showed us how self-replicating patterns could swirl, combine, divide, and evolve to develop the capability to understand all this.

* * *

At the beginning of this book, I said that I didn't want you to think that science is just about answers or that it's just about learning cut-and-dried facts. As I hope you've seen, despite the success that science and technology have had, much remains to be uncovered, including some very big questions. So perhaps the title *The Universe Explained with a Cookie* is a promise unfulfilled. Plenty in the universe is still unexplained.

But I hope this book has given you the appetite and confidence to explore these subjects more. Nothing beats that *aha* moment when something clicks into place.

Now go eat your cookie. You deserve it.

ACKNOWLEDGMENTS

The first person I need to thank is Daniel Fonder, my daughter's fifth-grade teacher. When I first had the idea to use a cookie to explain different science concepts, he graciously opened his classroom to allow me to come in and present to the students (badly), do experiments (that went wrong), and pass out cookies (which were enjoyed) to the class. Thanks also to my daughter for not being too mortified.

Thanks to Daniel Nayeri for bringing me into the Odd Dot family and giving me the opportunity to publish this book. My editor, Julia Sooy, has provided incredibly valuable feedback throughout the process, and the book would not be what it is today without her guidance.

The illustrations from Michael Korfhage are both delightful and charming, and it was a pleasure to work with him.

I'd like to thank my test readers Aimee Lam, Susan Engelstein, Bonnie Biel (Thanks, Mom!), Isaac Medford, and Paul Riggins. All your feedback was tremendous.

Finally, I would like to thank two authors whose philosophy has informed this book and my life—Douglas Hofstadter and Jacob Bronowski. I would not be who I am without their bold ideas.

RECOMMENDED READING

If you are interested in learning more about any of the
topics covered in *The Universe Explained with a Cookie*,
these books are highly recommended.

CHAPTER ONE: DARK MATTER EXPLAINED WITH FLOUR

Dark Matter and Dark Energy: The Hidden 95% of the Universe by Brian Clegg

CHAPTER TWO: FUSION EXPLAINED WITH SUGAR

Sun in a Bottle: The Strange History of Fusion and the Science of Wishful Thinking
  by Charles Seife

CHAPTER THREE: ATOMIC STRUCTURE EXPLAINED WITH SALT AND BAKING SODA

*The Disappearing Spoon: And Other True Tales of Madness, Love, and the History
  of the World from the Periodic Table of the Elements* by Sam Kean

CHAPTER FOUR: QUARKS EXPLAINED WITH A COOKIE SWAP

QED: The Strange Theory of Light and Matter by Richard Feynman

CHAPTER FIVE: QUANTUM MECHANICS EXPLAINED WITH MILK AND COOKIES

*Through Two Doors at Once: The Elegant Experiment That Captures the Enigma of
  Our Quantum Reality* by Anil Ananthaswamy

CHAPTER SIX: EVOLUTION EXPLAINED WITH BUTTER AND A BAKING COMPETITION

The Panda's Thumb by Stephen Jay Gould

What Evolution Is by Ernst Mayr

CHAPTER SEVEN: GENETIC ENGINEERING EXPLAINED WITH AN EGG

Genome: The Autobiography of a Species in 23 Chapters by Matt Ridley

The Code Breaker: Jennifer Doudna, Gene Editing, and the Future of the Human Race
  by Walter Isaacson

CHAPTER EIGHT: EMBRYONIC DEVELOPMENT EXPLAINED WITH COOKIE DECORATING

Endless Forms Most Beautiful: The New Science of Evo Devo by Sean B. Carroll

CHAPTER NINE: UNCERTAINTY EXPLAINED WITH ¾ CUP OF PACKED BROWN SUGAR

Naked Statistics: Stripping the Dread from the Data by Charles Wheelan

What is a p-value Anyway? 34 Stories to Help You Actually Understand Statistics by Andrew Vickers

CHAPTER TEN: THERMODYNAMICS EXPLAINED WITH BAKING AND AN ICE CREAM SANDWICH

A Matter of Degrees: What Temperature Reveals About the Past and Future of Our Species, Planet, and Universe by Gino Segre

CHAPTER ELEVEN: ENTROPY EXPLAINED WITH MIXING

Entropy: God's Dice Game by Oded Kafri and Hava Kafri

CHAPTER TWELVE: CHAOS EXPLAINED WITH VANILLA

Order Out of Chaos: Man's New Dialogue with Nature by Ilya Prigogine and Isabelle Stengers

The Order of Time by Carlo Rovelli

CHAPTER THIRTEEN: COMPLEXITY EXPLAINED WITH COOKIE CUTTERS

Complexity: The Emerging Science at the Edge of Order and Chaos by M. Mitchell Waldrop

CHAPTER FOURTEEN: FRACTALS EXPLAINED WITH AN OATMEAL RAISIN COOKIE

Gödel, Escher, Bach: an Eternal Golden Braid by Douglas Hofstadter

CHAPTER FIFTEEN: EXOPLANETS EXPLAINED WITH A NICE GOLDEN-BROWN COLOR

The Planet Factory: Exoplanets and the Search for a Second Earth by Elizabeth Tasker

CHAPTER SIXTEEN: THE BIG BANG EXPLAINED WITH CHOCOLATE CHIPS

The First Three Minutes: A Modern View of the Origin of the Universe by Steven Weinberg

The End of Everything (Astrophysically Speaking) by Katie Mack

INDEX